RANGELAND HEALTH

New Methods to Classify, Inventory,
and Monitor Rangelands

Committee on Rangeland Classification

Board on Agriculture

National Research Council

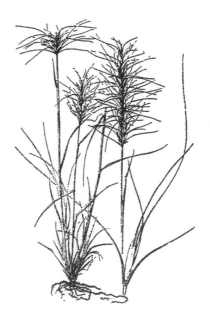

NATIONAL ACADEMY PRESS
Washington, D.C.   1994

NATIONAL ACADEMY PRESS • 2101 Constitution Avenue • Washington, D.C. 20418

NOTICE: The project that is the subject of this report was approved by the Governing Board of the National Research Council, whose members are drawn from the councils of the National Academy of Sciences, the National Academy of Engineering, and the Institute of Medicine. The members of the committee responsible for the report were chosen for their special competences and with regard for appropriate balance.

This report has been reviewed by a group other than the authors according to procedures approved by a Report Review Committee consisting of members of the National Academy of Sciences, the National Academy of Engineering, and the Institute of Medicine.

This material is based on work supported by the U.S. Department of Agriculture, Extension Service, under Agreement No. 90-EXCA-1-0102 and the Ford Foundation under Grant No. 880-0604. Additional funding was provided by the National Research Council. Dissemination was supported in part by the W. K. Kellogg Foundation.

Library of Congress Cataloging-in-Publication Data

Rangeland health : new methods to classify, inventory, and monitor
    rangelands / Committee on Rangeland Classification, Board on
    Agriculture, National Research Council.
        p.   cm.
    Includes bibliographical references (p.   ) and index.
    ISBN 0-309-04879-6
        1. Range management—United States. 2. Rangelands—United States.
    3. Range ecology—United States. 4. Range management.
    5. Rangelands. 6. Range ecology. I. National Research Council
    (U.S.). Committee on Rangeland Classification.
    SF85.3.R36   1993
    333.74'0973—dc20   93-39567
                                                                      CIP

Any opinions, findings, conclusions, or recommendations expressed in this publication are those of the author(s) and do not necessarily reflect the view of the organizations or agencies that provided support for this project.

Printed in the United States of America.

iii

v

Preface

Proper use and management of our nation's natural resources has been a challenge since European settlement of this country began. Rangelands and related resources were damaged during settlement by erosion and loss of habitats through inappropriate use and unintentional mismanagement.

During the 1890s through the 1930s, we as a society took actions to prevent further deterioration of our rangelands and to repair past damage. We placed large areas of rangeland in public trust and charged the U.S. Forest Service, the Bureau of Land Management, and other federal and state agencies to see that these lands are properly managed. We charged the Soil Conservation Service with the responsibility of helping private landowners improve management of their rangelands. We invested in research to improve our understanding of these complex ecosystems, established university programs to educate professional range managers, and established extension programs to provide continuing education programs for range users and agency personnel. Out of this developed a group of professionals—people and organizations dedicated to the proper management of the nation's rangelands.

Because livestock production was, for most of this century, the primary use of rangelands, range managers favored strategies, treatments, and methods that improved the range for livestock. Since World War II, and more particularly during the last quarter century, Americans have looked to rangelands as important areas for recreation and have become more concerned about the environmental condition of the country's rangelands. Issues such as riparian zones, wilderness areas, biological diversity, threatened and endangered species, including wild horses and burros, dominate much of the concerned public's interest in rangelands.

The current condition or quality of U.S. rangelands has been described by some as the best condition in this century and by others, using the

Thurber needlegrass (*Stipa thurburiana*)

same data, as extremely abused and degraded. Agencies have used different methods to evaluate the ecological condition of rangelands and have interpreted the data gathered using the same method in different ways. These different interpretations have confused the public, Congress, range users, management agencies, and range scientists themselves.

The Board on Agriculture of the National Research Council convened a committee in 1989 to examine the scientific basis of methods used by the Soil Conservation Service, Bureau of Land Management, and U.S. Forest Service to inventory, classify, and monitor rangelands. The Committee on Rangeland Classification Systems was charged to

• analyze current and historical procedures used by federal agencies to assess rangelands;
• assess the success of current systems as tools for characterizing rangeland health and ecological condition;
• identify the primary scientific obstacles to developing improved systems; and
• make recommendations for improving systems to better characterize the health of the nation's rangelands.

The committee completed its work through a series of meetings, field investigations, interviews with agency personnel, and discussions with other experts.

The committee concluded that a standard method and a common data base for evaluating rangelands is needed—one that everyone can understand and use to make decisions about use and management of and investment in our rangeland resources. We are hampered in our ability to make decisions and progress because of our inability to answer questions about the condition or quality of our rangelands. This report describes an approach for evaluating the ecological health of rangeland ecosystems.

The recommendations proposed by the committee are practical and applicable to the management of huge areas of land by people who have many, often conflicting, duties.

This report recommends that the principal purpose of rangeland assessments should be to assess rangeland health and recommends the criteria and methods that should be used to make that assessment. Chapter 1 explains why the health of the nation's rangelands should be of concern to the public, policymakers, scientists, and ranchers. Chapter 2 defines the committee's concept of rangeland health and the role assessments of rangeland health should play in managing rangelands. Chapter 3 analyzes the suitability of the methods used to assess rangelands as measures of rangeland health. Chapter 4 outlines the criteria that should be used to assess rangeland health. Chapter 5 details the inventorying and monitoring systems needed to make national-level assessments of the health of federal and nonfederal rangelands.

The strategy for evaluating rangeland health recommended in this report is a good first approximation of what is needed to do a better job of evaluating the ecological health of United States rangelands. The committee offers it to the profession of rangeland management and to society as a whole with this challenge: test it and change it, but do it in the same cooperative manner that this committee used to produce the strategy recommended in this report.

Frank E. "Fee" Busby, *Chair*
Committee on Rangeland Classification

Acknowledgments

The committee acknowledges the contributions of several individuals who helped prepare this report.

Richard Wiles conceived and initiated this study and directed the committee's work during its first 2 years. The direction, support, and vision he provided during the early part of this study is an important reason for its success. Dr. Thadis W. Box chaired the planning meeting that preceded the initiation of this study. He and all the participants at the planning meeting helped define the topics, objectives, and methods that were used by the committee in completing this report.

Numerous individuals within the U.S. Department of Agriculture's Soil Conservation Service and Forest Service and the U.S. Department of the Interior's Bureau of Land Management provided data, prepared presentations, organized field trips, and explained agency methods for the committee. The cooperation of this dedicated group of professionals was essential to the completion of this study.

The committee also acknowledges the work of the staff of the Board on Agriculture. Craig Cox, Project Director, provided invaluable advice and direction during the final phase of this study. His energy and persistence helped the committee through a long, and sometimes difficult, process of reaching consensus. Cris Banks, Administrative Assistant, provided gracious and effective support, as did her predecessors, Suzanne Mason and Amy Gorena. Finally, Carla Carlson, Director of Communications; Janet Overton, Editor; and Michael Hayes, Editorial Consultant, provided editorial support. Their insistence on readability and quality are evident on every page of this report.

Rocky Mountain juniper (*Juniperus scopulorum*)

The National Academy of Sciences is a private, nonprofit, self-perpetuating society of distinguished scholars engaged in scientific and engineering research, dedicated to the furtherance of science and technology and to their use for the general welfare. Upon the authority of the charter granted to it by the Congress in 1863, the Academy has a mandate that requires it to advise the federal government on scientific and technical matters. Dr. Bruce M. Alberts is president of the National Academy of Sciences.

The National Academy of Engineering was established in 1964, under the charter of the National Academy of Sciences, as a parallel organization of outstanding engineers. It is autonomous in its administration and in the selection of its members, sharing with the National Academy of Sciences the responsibility for advising the federal government. The National Academy of Engineering also sponsors engineering programs aimed at meeting national needs, encourages education and research, and recognizes the superior achievements of engineers. Dr. Robert M. White is president of the National Academy of Engineering.

The Institute of Medicine was established in 1970 by the National Academy of Sciences to secure the services of eminent members of appropriate professions in the examination of policy matters pertaining to the health of the public. The Institute acts under the responsibility given to the National Academy of Sciences by its congressional charter to be an adviser to the federal government and, upon its own initiative, to identify issues of medical care, research, and education. Dr. Kenneth I. Shine is president of the Institute of Medicine.

The National Research Council was organized by the National Academy of Sciences in 1916 to associate the broad community of science and technology with the Academy's purposes of furthering knowledge and advising the federal government. Functioning in accordance with general policies determined by the Academy, the Council has become the principal operating agency of both the National Academy of Sciences and the National Academy of Engineering in providing services to the government, the public, and the scientific and engineering communities. The Council is administered jointly by both Academies and the Institute of Medicine. Dr. Bruce M. Alberts and Dr. Robert M. White are chairman and vice-chairman, respectively, of the National Research Council.

Contents

Indian ricegrass (*Oryzopsis hymenoides*)

5   INVENTORYING AND MONITORING RANGELAND HEALTH 134
Past Inventories of Rangelands, 134
Current Inventorying and Monitoring Systems, 146
National System of Inventorying and Monitoring
     Rangeland Health Is Needed, 151
Notes, 156

 # Tables and Figures

TABLES

FIGURES

Crested wheatgrass (*Agropyron cristatum*)

RANGELAND HEALTH

Executive Summary

The United States has about 312 million hectares (770 million acres) of rangelands, from the wet grasslands of Florida to the desert shrub ecosystems of Wyoming and from the high mountain meadows of Utah to the desert floor of California. These diverse ecosystems produce an equally diverse array of tangible and intangible products. Commodities, such as forage for livestock, wildlife habitat, water, minerals, energy, recreational opportunities, some wood products, and plant and animal genes, are important economic goods. Rangelands also produce intangible products such as natural beauty and wilderness that satisfy important societal values and that can be as economically important as more tangible commodities. More than half of these rangelands are privately owned, 43 percent are owned by the federal government, and the remainder are owned by state and local governments.

STATE OF RANGELAND ECOSYSTEMS

Rangeland degradation reduces the diversity and amount of the values and commodities that rangelands provide, and severe rangeland degradation can be irreversible. Overgrazing, drought, erosion, and other human and naturally induced stresses have caused severe degradation in the past. Although most observers agree that rangeland degradation was widespread on overgrazed and drought-plagued rangelands at the turn of the century, the present state of health of U.S. rangelands is a matter of sharp debate. Added to that debate is the confusion caused by each agency using individual agency-specific terminology to identify conditions and methods. To lessen that confusion, in this report terms will be followed by the initials of the agency to which that term is specific.

Big sagebrush (*Artemisia tridentata*)

1

An example of severe water erosion, gullies on this hillside are the result of flash flooding after a torrential rainfall on overgrazed sheep range. Credit: U.S. Department of Agriculture.

Questions about Methods

Range condition [Soil Conservation Service (SCS)] and ecological status [U.S. Forest Service (USFS) and Bureau of Land Management (BLM)] assessments have been the primary methods used to evaluate rangelands and have, as recently as 1989, been interpreted as measures of rangeland health. Now, however, the scientific debate over the use of current methods to assess rangeland ecosystems has intensified, leading to disagreements over the proper interpretation of past and ongoing range condition (SCS) and ecological status (USFS and BLM) assessments. Different scientists looking at the same data have come to different conclusions about both the state of U.S. rangelands and the value of the available data.

Lack of Reliable Data

All existing national-level rangeland assessments suffer from the lack of current, comprehensive, and statistically representative data obtained in the field. Data collected by the same methods over time and by a sampling design that allows aggregation of the data at the national level are not available for assessing federal and nonfederal rangelands. The data that are available for assessing the status of rangelands have been obtained by many different methods and from many different sources.

Need to Know Is Urgent

The debate and uncertainty over the health of the nation's rangelands have become inextricably bound to the debate over the best use and management of federal rangelands administered by BLM, USFS, and other agencies of the federal government. Similar concerns about the management of nonfederal rangelands have been voiced, particularly over the effect of grazing and other uses, on water quality. The fact that the available data do not allow investigators to reach definitive conclusions about the state of rangelands seriously impedes efforts to resolve the debate about proper management of the nation's federal and nonfederal rangelands.

There is an urgent need to develop the methods and data collection systems at both the local and national levels to assess federal and nonfederal rangelands. The importance of the values and commodities provided by rangelands, the history of rangeland degradation, the evidence pointing to ongoing rangeland degradation, and the inadequate data on current conditions at both the local and the national levels suggest that it is unwise to neglect the nation's federal and nonfederal rangelands.

PURPOSES OF NATIONAL ASSESSMENTS

The choice of methods and criteria to assess rangelands entails a judgment about what information is most important to provide national policymakers, ranchers, environmentalists, and the general public about rangelands. National assessments could be designed to answer many different questions about rangelands. Moreover, deciding which information is most important depends, in large part, on the relative importance society places on the various values and commodities rangelands provide.

The importance of providing the information needed to protect and sustain the capacity of rangeland ecosystems to provide the values and commodities desired by society has been repeatedly recognized in national legislation. SCS, USFS, and BLM have been mandated to provide the assessments of rangeland ecosystems needed to protect the quality and sustained yield of renewable resources. Providing policymakers and the public with the information needed to determine whether the capacity of rangelands to satisfy values and produce commodities is being sustained, improved, or degraded should be the primary goal of national assessments of rangelands.

Standards for National Assessments

Divergent views on the proper interpretation of current rangeland classification and inventory methods have confused the debate over proper rangeland management. An agreed-to standard that can be used to determine whether the capacity of these rangelands to produce commodities and satisfy values is being conserved, degraded, or improved is needed. The lack of a consistently defined standard for acceptable conditions of rangeland ecosystems is the most significant limitation to current efforts to assess rangelands. The lack of such an agreed-to standard has and continues to confuse the public, the U.S. Congress, ranchers, and range scientists themselves.

DEFINITION OF RANGELAND HEALTH

Rangeland health should be defined as the degree to which the integrity of the soil and the ecological processes of rangeland ecosystems are sustained.

Rangelands are ecosystems, not individual organisms, and the use of the term "health" should not imply that simple analogies can be made between the health of an organism and the health of an ecosystem. Health, however, has been used to indicate the proper functioning of complex systems; the term is increasingly applied to ecosystems to indicate a condition in which ecological processes are functioning properly to maintain the structure, organization, and activity of the system over time.

The capacity of rangelands to produce commodities and to satisfy values on a sustained basis depends on internal, self-sustaining ecological processes such as soil development, nutrient cycling, energy flow, and the structure and dynamics of plant and animal communities. *Webster's Third New International Dictionary* defines *healthy* as (1) functioning properly or normally in its vital functions, (2) free from malfunctioning of any kind, and (3) productive of good of any kind. The terms "healthy" or "unhealthy" are most properly applied to ecosystems as an indication of proper or normal functioning of ecological processes resulting in the production of goods, that is, commodities or values. More specifically, the committee recommends the term "rangeland health" be used to indicate the degree of integrity of the soil and ecological processes that are most important in sustaining the capacity of rangelands to satisfy values and produce commodities.

MINIMUM ECOLOGICAL STANDARD

The minimum standard for rangeland management should be to prevent human-induced loss of rangeland health.
Large investments of time, money, and energy are required to restore unhealthy rangelands. Even with restoration, there may be permanent loss of capacity to produce commodities and satisfy values or loss of options to use and manage those rangelands in the future. Any human-induced loss of rangeland health should be prevented.

In other agricultural ecosystems, such as intensively managed croplands, the capacity to satisfy values and produce commodities is often augmented by using high levels of external inputs such as irrigation water or fertilizer, the physical environment is modified by tillage or terracing of the land, and pests are controlled by applying chemical pesticides. Rangelands, for the most part, do not receive such inputs. The capacity of rangelands to produce commodities and satisfy values depends on the integrity of internal nutrient cycles; energy flows; plant community dynamics; an intact soil profile; and stores of nutrients and water.

Rangeland health should be a minimum ecological standard, independent of the rangeland's use and how it is managed. If rangeland health is protected, a variety of uses could be appropriate for any particular rangeland. The selected use(s) would depend on the preferences of the landowner, if the rangeland is privately owned, or the relative values placed by society on different uses of federal rangelands. These decisions will still be contentious, but they can at least be taken in the context of protecting rangeland health and, therefore, the capacity of rangelands to produce commodities and satisfy values regardless of their use.

USE AND MANAGEMENT

Rangeland health inventories and monitoring systems should be one part of a larger system of data gathering and analysis to inform range mangers, policymakers, and the public about the use and management of rangelands.
An assessment of rangeland health estimates the risk of the loss of the capacity to produce commodities and satisfy values by evaluating the integrity of the site's ecological processes and soils. Such an evaluation

does not determine conclusively the causes of current conditions or determine what changes in management are required, or how a particular rangeland should be used. The determination of which uses and management practices are appropriate will require the evaluation of different data. No single index will meet all the needs of rangeland assessment and management.

Categories for National Assessments

The principal purpose of rangeland inventories and assessments completed by the Soil Conservation Service (SCS) of the U.S. Department of Agriculture (USDA), the Bureau of Land Management (BLM) of the U.S. Department of the Interior (DOI), and the U.S. Forest Service (USFS) of USDA should be to determine the location and proportion of rangelands that are healthy, at risk, or unhealthy.

Rangeland ecosystems are dynamic systems that are constantly adjusting to changes in the environment. Fitting rangelands into categories based on assessments of ecological health is a difficult but an essential task of national assessments. The categories defined for purposes of national rangeland assessments should facilitate the interpretation of the results of those assessments for policymakers, range managers, ranchers, and the public. The categories used for national assessments should signal where rangeland use and management need to be changed and help direct range management and technical assistance to those rangelands where they are most needed to prevent degradation or to improve damaged rangelands.

The committee recommends that rangelands be placed in three broad categories based on an evaluation of the soil and ecological processes. Rangelands should be considered (1) healthy if an evaluation of the soil and ecological processes indicate that the capacity to satisfy values and produce commodities is being sustained; (2) at risk if the assessment of current conditions indicates a reversible loss in productive capacity and increased vulnerability to irreversible degradation; and (3) unhealthy if the assessment indicates that degradation has resulted in loss of capacity to provide values and commodities that cannot be reversed without external inputs.

METHODS TO ASSESS RANGELAND HEALTH

The evaluation of rangeland health will require the collection of additional and different data and new approaches to interpreting those data. These data and approaches should reflect the diverse processes of rangeland ecosystems that sustain their capacity to satisfy values and produce

A healthy rangeland has the sustained capacity to satisfy values and produce commodities. Credit: USDA U.S. Forestry Service.

commodities. Soil stability, watershed function, nutrient cycling, energy flow, and the mechanisms that enable recovery from stress should be assessed on rangelands. Established criteria are needed to determine, based on the suite of indicators that are sampled, whether the ecosystem is healthy, at risk, or unhealthy.

Defining Boundaries

Categorizing rangelands as healthy, at risk, or unhealthy requires defining two boundaries: the boundary distinguishing healthy from at-risk rangelands and the boundary distinguishing at-risk from unhealthy rangelands. Rangelands adapt to changes in their use or management and in the environment through alterations in ecosystem characteristics such as plant composition, the amount of plant biomass produced, the amount of nutrients and the rate at which they are cycled, and the amount and composition of soil organic matter. The ecological state of the rangeland at a point in time is the sum total of these characteristics. The rangeland ecosystem shifts between different ecological states over time in response to natural or human-induced factors. Such changes can be sudden or they may occur gradually.

There are important differences between the processes and effects of change, however, that can be used to identify boundaries between healthy, at risk, and unhealthy rangelands for the purposes of national assessments. Many changes in ecological state may have no long-term effect on the capacity of the rangeland to produce commodities or satisfy values. Other changes can be destructive, but their destructive effects can be reversed by changes in use and management or as natural conditions improve. Some changes, however, such as serious soil degradation, the interruption of nutrient cycles, and the loss of important species or seed sources can lead to irreversible changes that reduce the amount and diversity of vegetation, habitat, aesthetics, and other commodities and values the rangeland can provide even if use and management or natural conditions improve. The boundaries between healthy, at-risk, and unhealthy states of a rangeland ecosystem should be distinguished based on differences between states in the capacity to produce commodities and satisfy values and on the reversibility of the changes between states.

The *threshold of rangeland health* should define the boundary between unhealthy and at-risk states. The threshold of rangeland health should be distinguished from other boundaries between the ecological states of a rangeland ecosystem by two key factors: (1) the shift from the at-risk to the unhealthy state is not easily reversed, and (2) the change from the at-risk to unhealthy state entails a reduction in the capacity of the rangeland to satisfy values or produce commodities. The *early warning line* should define the boundary between healthy and at-risk states. The shift between healthy and at-risk states should indicate: (1) a loss in capacity to satisfy values and produce commodities that is reversible through change in use or management and (2) an increased vulnerability to irreversible degradation. An at-risk categorization should signal the need to take corrective action or to further investigate the site to determine the seriousness and causes of the degradation.

Criteria and Indicators

The determination of whether a rangeland is healthy, at risk, or unhealthy should be based on the evaluation of three criteria: degree of soil stability and watershed function, integrity of nutrient cycles and energy flows, and presence of functioning recovery mechanisms.

Judgments based on the preponderance of evidence from the evaluation of multiple criteria will be required for meaningful assessments of rangeland health. No single criterion alone will be sufficient to determine whether rangelands are healthy, at risk, or unhealthy.

SOIL STABILITY AND WATERSHED FUNCTION

Soil degradation, primarily through accelerated erosion by wind and water, causes a direct and often irreversible loss of rangeland health. Soil degradation not only damages the soil itself but also disrupts nutrient cycling, water infiltration, seed germination, seedling development, and other ecological processes that are important components of rangeland ecosystems. In addition, soil degradation damages watersheds, which leads to further degradation of rangeland ecosystems as well as water pollution. Indicators of soil stability and watershed function should be central to the evaluation of rangeland health.

The indicators selected to assess soil stability and watershed function should relate to two fundamental processes: (1) soil erosion by wind and water and (2) infiltration or capture of precipitation. The development of predictive models that estimate rates of soil loss and infiltration coupled with the establishment of acceptable rates of soil erosion could help to quantify soil stability and watershed function for an evaluation of rangeland health. Reliable predictive models are being developed but do not yet exist for rangelands.

Soil surface characteristics are currently the best available indicators of soil stability and watershed function. Soil surface characteristics, such as presence of rills and gullies or pedestaling of plants, have been widely used as indicators of the degree of soil movement and the condition of the soil surface. Soil surface characteristics also give partial evidence of the magnitude of infiltration or runoff from a site. An evaluation of soil stability and watershed function, as determined by the use of soil surface characteristics as indicators of soil erosion and runoff, should become a fundamental component of all inventorying and monitoring programs for federal and nonfederal rangelands.

NUTRIENT CYCLING AND ENERGY FLOW

The capacity of rangelands to produce commodities and satisfy values depends on the capture of sunlight energy through photosynthesis and on the accumulation and cycling of nutrients over time. Interruption or slowing of nutrient cycling or energy flows in time or space can lead to degradation as a rangeland site becomes increasingly lacking in available nutrients, energy, and biomass.

Plants depend on the nutrients in the soil and energy captured from the sun. Nutrients stored in the soil are used and reused by plants, animals, and microorganisms. The amount of nutrients available and the speed with which nutrients cycle between plants, animals, and the soil are fundamental components of rangeland health. Similarly, the amount, timing, and distribution of energy captured through photosynthesis are fundamental to the function of rangeland ecosystems. The total amount of energy captured from sunlight is an important determinant of the commodities produced and values satisfied by rangelands. Indicators that

Only if rangeland health is sustained or improved can the debate profitably shift to whether rangelands are suitable for the production of livestock, wildlife, recreation, or some combination of these uses. Credit: USDA U.S. Forest Service.

can be used to evaluate the integrity of nutrient cycles and energy flows should be part of a comprehensive evaluation of rangeland health. Experience with such indicators is limited, but the degree of fragmentation in the distribution of plants, litter, roots, and photosynthetic period may be useful starting points for the development of more quantitative indicators of nutrient cycles and energy flow.

RECOVERY MECHANISMS

The capacity of rangeland ecosystems to adjust to change in ways that prevent loss of rangeland health depends on the presence or absence of functioning recovery mechanisms. Properly functioning recovery mechanisms result in the capture and cycling of nutrients; capture of energy; conservation of nutrients, energy, and water within the site; development of resistance to extreme events; and resilience to change—the processes through which rangeland health is sustained or improved.

Indicators of change in recovery mechanisms should be part of a comprehensive assessment of rangeland health. Useful indicators may include increasing vegetative cover, increasing plant vigor, change in the kind and number of seedlings, changes in plant age class distribution, and other plant community attributes that will lead to greater soil stability, nutrient storage and cycling, and energy capture. Various indicators of plant demographics have been commonly used in rangeland assessments, and indicators of age class distribution, plant vigor, and the presence and distribution of microsites for seed germination and seedling development would be useful starting points for the development of more systematic indicators of the function of recovery mechanisms on rangelands.

Research Needed

The secretaries of USDA and DOI should initiate a coordinated research effort, drawing on federal agency and other scientists to develop, test, and employ indicators of the spatial and temporal distributions of nutrients and energy and the presence and functioning of recovery mechanisms for use in rangeland health assessments.

The lack of experience with and testing of specific indicators of nutrient cycling, energy flow, and recovery mechanisms is an important impediment to the development of a comprehensive system of determining whether rangelands are healthy, at risk, or unhealthy. There is an urgent need for basic and applied research to develop useful indicators and the understanding needed to interpret the significance of changes in those indicators.

The secretaries of USDA and DOI should initiate a coordinated research effort, drawing on federal agency and other scientists to develop, test, and employ new models of rangeland change that incorporate the concept of ecological thresholds.

New models that better explain the dynamics of rangeland ecosystems are needed to provide the foundation for rangeland health assessments. New models have been proposed, but as yet there is no single, coherent model that explains the anomalies in the current succession-retrogression model or that has been sufficiently tested to replace current successional concepts. An interdisciplinary research effort that links range scientists with other ecologists is needed to develop and test new models.

The secretaries of USDA and DOI should initiate a coordinated research effort, drawing on federal agency and other scientists to increase understanding of the relationship between soil properties and rangeland health.

While much research and experience supports the relationship of soil surface characteristics to rangeland health, basic knowledge of the effects of other soil properties such as organic matter content or water-holding capacity on nutrient cycling, energy flows, recovery mechanisms, and other elements of rangeland health is limited. The effects of grazing management and other management practices on soil properties are also not well understood. Basic and applied research is needed to increase understanding of how changes in soil properties affect rangeland health.

NATIONAL INVENTORYING AND MONITORING SYSTEM

An assessment of the health of a particular rangeland will provide information to local managers, ranchers, and others who need to protect or improve that rangeland. A coordinated system that can be used to complete national-level assessments of rangeland health, however, is not in place. The lack of a national-level inventorying and monitoring system is a major impediment to the nation's ability to assess the health of federal and nonfederal rangelands and to judge whether current management and use of federal and nonfederal rangelands are adequately sustaining the rangeland's capacity to satisfy values and produce commodities.

Minimum Data Set

A national system to inventory and monitor rangelands should be based on the collection and analysis of data on changes in a minimum set of multiple indicators of rangeland health.

Rangelands are diverse, and large amounts of data are needed for all range management activities. Some indicators will be of particular importance for some types of rangelands and for some management purposes. Indicator species of plants or animals, for example, could be particularly important for specific rangeland ecosystems. To build the consistent data set required to assess rangeland health, a small, selected set of indicators should be collected as part of all current and ongoing rangeland management and assessment activities on both federal and nonfederal rangelands. This minimum data set can be augmented with measures of additional indicators of rangeland health that are of particular importance for the assessment of particular classes of rangelands.

Standardize Indicators and Methods

The secretaries of USDA and DOI should convene a multiagency task force to develop, test, and standardize indicators and methods for inventorying and monitoring rangeland health on federal and nonfederal rangelands.

Current discrepancies in the definitions, interpretations, and methods have seriously reduced the comparability as well as the utility of the data collected by SCS, BLM, and USFS. Similarly, new methods of rangeland assessments developed by SCS, BLM, and USFS should be coordinated with efforts of the Environmental Protection Agency (EPA) to develop the Environmental Monitoring and Assessment Program (EMAP). The same discrepancy problems will plague efforts to make national-level assessments of rangeland health if SCS, BLM, USFS, and EPA independently develop different methods.

The multiagency task force should coordinate federal efforts, including EPA's EMAP effort, leading to (1) a set of indicators that should be included in a minimum data set for inventorying and monitoring rangeland health; (2) standard methods of measuring indicators and categorizing rangelands as healthy, at risk, or unhealthy; (3) a series of field tests to validate the indicators and methods selected; and (4) quantification of the correlation between measures of rangeland health and range condition (SCS) or ecological status (USFS and BLM).

It is also important that all agencies adopt comparable systems of site classification for the purposes of national-level inventorying and monitoring of rangelands. To limit differences in interpretations, common site classifications should be soil based and should provide general information on vegetative production, plant and animal community structure, life-form dynamics, and predicted responses to disturbances such as fires, grazing, floods, and droughts. Correlation of rangeland classifications across administrative boundaries will be needed even if all agencies adopt unified approaches to site classification for inventorying and monitoring purposes.

National Sampling System

The secretaries of USDA and DOI should develop coordinated plans for implementing a sampling system on federal and nonfederal rangelands that will produce estimates of the proportion of healthy, at-risk, and unhealthy rangelands that are significant at appropriate local, state, regional, and national levels.

A national sampling system that coordinates the activities of USDA, DOI, and EPA is needed to collect, analyze, and aggregate data to determine the proportion of federal and nonfederal rangelands that are healthy, at risk, or unhealthy. The National Resources Inventory, conducted by SCS, is a statistically valid sampling design used to assess various characteristics of nonfederal rangelands. No comparable sampling program is in place on federal rangelands. Most of the data collected are for management purposes rather than for national inventorying and monitoring purposes. The development of a coordinated sampling system for both federal and nonfederal rangelands is urgently needed.

Periodic Sampling Needed

The secretaries of USDA and DOI should develop coordinated plans for implementing periodic sampling of federal and nonfederal rangelands to determine changes in the proportions of healthy, at-risk, and unhealthy rangelands.
Periodic monitoring must be a fundamental part of a valid national system for evaluating rangeland health. The periodicity of repeated sampling should reflect the rapidity of change within the indicators selected to monitor rangelands and the degree of degradation that a change implies to give adequate early warning of increases in the area of unhealthy rangelands. Monitoring should be frequent enough such that a rangeland would not slip from a healthy to an unhealthy state between sampling periods.

TRANSITION TO RANGELAND HEALTH ASSESSMENTS

The implementation of rangeland health assessment as a management tool to protect rangelands at the local level or as a national inventory and monitoring program will take time. Standardized indicators and methods will have to be developed. National sampling systems will have to be put in place, and resources will have to be allocated to the collection of data on indicators of rangeland health. Given the importance of rangeland ecosystems, it is important that SCS, USFS, and BLM move quickly and in a coordinated fashion toward developing and employing the tools needed to assess rangeland health.

Intermediate Steps

As progress is made toward more comprehensive and standardized assessments of rangelands, there are important intermediate steps that could be taken to substantially increase the information available and the understanding needed to determine whether rangelands are healthy, at risk, or unhealthy.

Indicators of soil surface condition should be added to all current and ongoing range condition (SCS) and ecological status (USFS and BLM) assessments, and any other ongoing efforts to assess rangelands, as a first step toward a more comprehensive evaluation of rangeland health.
There is much experience with the use of soil surface characteristics as indicators of soil stability and watershed function. The addition of indicators of soil surface condition to all current and ongoing efforts to assess rangelands would be a useful first step toward a more comprehensive system of evaluating rangeland health—a step that should be taken

immediately. Data on the soil surface condition of rangelands should also be collected as part of the National Resources Inventory.

All current and ongoing rangeland assessments done as part of Resources Conservation Act (RCA) appraisals, Resources Planning Act (RPA) assessments, national forest planning, USFS and BLM land use and allotment planning, and environmental assessments should be based on the analysis of multiple ecological attributes.

SCS, USFS, and BLM should analyze multiple ecological attributes of rangelands as part of current rangeland assessments and appraisals. Plant composition and production data collected as part of range condition (SCS) and ecological status (USFS and BLM) ratings should be analyzed in conjunction with information collected on indicators of soil surface condition, as recommended above, and all other available information on erosion rates. Using these multiple indicators, the agencies could begin to assess soil stability and watershed function, distribution of nutrients and energy, and presence of functioning recovery mechanisms as a means of identifying rangelands at greater risk of loss of health. This analysis should be part of conservation planning or management of grazing allotments as well as national appraisals and assessments.

These assessments would not provide a complete assessment of rangeland health, but they would represent progress toward measuring and analyzing multiple ecological attributes within each agency. They would also help guide national policy for managing federal and nonfederal rangelands in the interim while more comprehensive and systematic assessments of rangeland health are developed.

Basic data including soil surface conditions, erosion rates, plant composition, and biomass production assembled and used to assess rangelands as part of RCA appraisals, RPA assessments, national forest planning, environmental assessments, and other assessments of federal and nonfederal rangelands should be made available to the public and the scientific community for independent review.

Independent review and analysis of these data will increase the understanding of and confidence in the results of the assessments of federal and nonfederal rangelands. Publication of basic data will provide a data set for scientific evaluation of the utility of alternative indicators of soil stability and watershed function, distribution of nutrients and energy, and presence of recovery mechanisms as measures of rangeland health. The availability of basic data, for example, used to estimate erosion rates as part of the National Resources Inventory has allowed scientists to test the effect of alternative agricultural policies and crop management practices on erosion rates. Independent review of these basic data has increased the confidence of estimates of erosion reductions expected from changes in farming practices. It is important that basic data on multiple ecological attributes of federal and nonfederal rangelands be made available to both the public and the scientific community to accelerate the transition to comprehensive methods for assessing rangeland health.

Preserving Continuity During the Transition

It is essential that, as the transition to more comprehensive assessments of rangeland health is made, a link be maintained between current methods and data and new methods and data. Current methods should not be abandoned until new systems are in place to replace or augment those methods. In most cases, the data collected as part of range condition (SCS) and ecological status (USFS and BLM) ratings are the only historical data available on rangelands. Although these data alone are not sufficient for assessing the health of the nation's rangelands, the continuity of these data should be preserved as the transition to rangeland health assessments is made.

SCS, USFS, and BLM should continue current and ongoing range condition (SCS) and ecological status (USFS and BLM) ratings while the transition to rangeland health assessment is made.

The data that have been and continue to be collected for range condition (SCS) and ecological status (USFS and BLM) assessments should provide a critical historical data set for use in judging changes in rangeland conditions. As a transition is made to national-level inventorying and monitoring of rangeland health as recommended here, it is imperative that this information not be lost. The committee strongly recommends that current and planned monitoring efforts that use range condition (SCS) and ecological status (USFS and BLM) move ahead and be augmented by the collection of additional data for evaluating rangeland health.

CHALLENGE TO RANGE SCIENTISTS AND MANAGERS

It will be difficult to develop methods of rangeland health assessment that are suitable for use by range managers who must administer large areas of federal rangeland or deliver technical assistance to ranchers and by scientists who develop national-level inventories of rangelands. New partnerships between range managers, range scientists, and other ecologists working in different ecosystems and different institutions will be needed. The barriers to coordination between different federal agencies erected by different mandates and traditions will have to be overcome.

Answering the question "Are our rangelands healthy?" may be the most important contribution range scientists and managers can make to resolving the debate over the use and management of federal and nonfederal rangelands. Answering this question will provide the information that is urgently needed by range managers, scientists, policymakers, ranchers, and environmentalists struggling to improve rangelands and range management. Advances in the methods needed to answer this question will help build a firmer scientific foundation for rangeland assessment and management. And finding ways to answer the question "Are our rangelands healthy?" will be an important step toward sustaining the ecological integrity and productivity of these important ecosystems.

1 Rangelands Are Important

In the United States, including Alaska, there are about 312 million hectares (770 million acres) of rangelands (U.S. Department of Agriculture, U.S. Forest Service, 1989a). This large area spans vastly different landforms and climates. Within these landforms are a remarkable diversity of rangeland ecosystems, from the wet grasslands of Florida to the desert shrub ecosystems of Wyoming and from the high mountain meadows of Utah to the desert floor of California. These diverse ecosystems provide a wide array of tangible commodities and values for society.

More than half of the nation's rangelands are privately owned, 43 percent are owned by the federal government, and the remainder are owned by state and local governments (Joyce [1989], as cited by Box [1990]) (Figure 1-1). At least 110 million hectares (272 million acres) of the rangelands present at the time of European settlement in the coterminous United States have been converted from rangelands to croplands, forests, urban areas, industrial sites, highways, and reservoirs (Klopatek et al., 1979). Although some eastern states such as Florida still have large areas of native rangelands, most of the remaining rangelands are found in 17 western states and Alaska.

RANGELAND MANAGEMENT AND USES

The western rangelands are the legendary wide open spaces of American history and mythology. Federal rangelands are managed chiefly by the Bureau of Land Management (BLM), an agency of the U.S. Department of the Interior (DOI), and the U.S. Forest Service (USFS), an agency of the U.S. Department of Agriculture (USDA). Federal and nonfederal rangelands include deserts, grasslands, canyons, tundra, mountains, and riparian areas (the grassy or woody areas located on the banks of a natural

Prairie June grass (*Koelaria cristata*)

18

watercourse, such as a river, lake, or tidewater). They include wilderness areas and provide habitat for millions of wild animals, plants, and fish, including 74 threatened or endangered species alone on lands administered by BLM (W. H. Radtkey, Bureau of Land Management, personal communication, 1992). They are increasingly used as an immense recreational resource by millions of visitors each year.

Whether publicly or privately owned, rangelands produce tangible products such as forage, wildlife habitat, water, minerals, energy, plant and animal gene pools, recreational opportunities, and some wood products. The chief commercial use of rangelands in the United States—and most of the world—is livestock grazing to produce food, fiber, and draft animals. These are referred to as *commodities* in this volume. Rangelands also produce intangible products (referred to as *values*) such as natural beauty, open spaces, and the opportunity for the ecological study of natural ecosystems.

Grazing lands in the United States include rangelands, forests, and pastures. Federal and nonfederal lands produced some 399,567,000 animal-unit months (the amount of forage consumed by an animal unit, usually estimated at 363 kilograms [800 pounds], in 1 month) of forage for beef cattle and sheep in 1985; federal lands produced 7 percent and non-

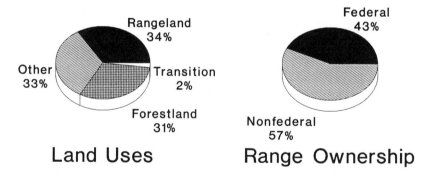

Land Uses   Range Ownership

FIGURE 1-1   Pie charts show percent of total U.S. land use and percent of rangeland ownership. Exact numbers are difficult to determine because statistical sources often define rangeland to include natural grasslands, savannahs, wetlands, deserts, and tundra. In addition, transition land includes land characteristics for which the predominant vegetation is grass or forage plants used for grazing. Also, although it is noted that more than 50 percent is nonfederal land, it is not clear exactly how much is owned by state and local governments (approximately 5 percent of nonfederal land is owned by state and local governments). Another problem in making exact determinations is that the system of deeding and rights use is so complicated, it is sometimes not entirely clear where government ownership ends and private ownership begins.

federal lands produced 93 percent of the total animal-unit months of forage consumed from rangelands, forestlands, and pasturelands (Gee et al., 1992).

Rangeland watersheds are important regulators of the quantity and quality of water in streams, lakes, and aquifers (an aquifer is a water-bearing layer of permeable rock, sand, or gravel beneath the earth's surface). Management of rangeland watersheds to increase the amount of clean water available for use by irrigators, municipalities, and industry and for recreational purposes is increasingly important. Federal and non-federal rangelands provide grazing areas for wild herbivores such as deer, antelope, and elk. Many species of fish and wildlife depend on rangelands and their associated streams and lakes for habitat. During some part of the year, rangeland ecosystems are associated with 84 and 74 percent of the total number of mammalian and avian species, respectively, found in the United States (Flather and Hoekstra, 1989).

Human Interactions on Western U.S. Rangelands

Soon after explorers discovered the coastlines of the Americas in 1540, they traveled inland, migrating north from Mexico and, later, west from New England. The growth of the population and commerce in what is now the western United States occurred over what can be viewed as three periods of human interaction with the land:

- the introduction of livestock on the open range 450 years ago,
- the spread of crop farming until the Dust Bowl in the 1930s, and
- the increased attention to the land as a recreational and aesthetic resource during a time of increasing urbanization.

Each change in the perceived value of the range brought with it changing ecological concerns.

In 1540, the Spanish explorer Francisco Vasquez de Coronado introduced the first domestic livestock to the open range of what is now the southwestern United States: 500 cattle, 5,000 sheep, and 1,000 horses (Wallace, 1936). As Spanish missionaries established and fostered outposts in the 1700s in areas that are now Texas, New Mexico, Arizona, and California, they brought an estimated 50,000 sheep and 20,000 cattle north from Mexico (Wallace, 1936).

The Mexican government liberally granted rangelands to people interested in establishing ranches in what is now the southwestern United States. By 1860, the number of cattle in California reached an estimated 3.5 million head (Burcham, 1961), and in Texas, the herd population soared from 330,000 in 1850 to some 4 million in 1860 (Paul, 1988). Access

Rangelands offer many recreational opportunities, including hiking, horseback riding, picnicking, skiing, fishing, hunting, snowmobiling, and driving of off-road vehicles. The demand for rangelands for recreational purposes is growing; for example, the demand for horseback riding is expected to double by the year 2040 (Cordell, 1989). The fees charged for the recreational use of privately owned rangelands are growing sources of revenue for rangeland owners (Box, 1990). The value placed on the recreational opportunities and open space provided by rangelands is expected to increase with increased levels of urbanization (Joyce, 1989).

CONCERN ABOUT THE STATE OF U.S. RANGELANDS

Conditions on U.S. rangelands have long been a source of concern. The Europeans who brought sheep and cattle to the rangelands of the western United States overestimated the ability of the land to support

---

to railroads in Sedalia, Missouri, and Abilene, Kansas, encouraged the beef boom.

Severe flooding in 1862 in California followed by 2 years of intense drought reduced the herds by between 200,000 and 1 million head (Burcham, 1961). California ranchers turned their attention to sheep, which they hoped would be better suited to the weather conditions of western rangelands, but this caused a debate among people who used rangelands, with cattle ranchers contending that sheep depleted all palatable grasses. Weather problems such as drought, blizzards, and storms plagued the rest of the southwestern range in the late 1880s, devastating cattle production. Prices for cattle in Chicago stockyards dropped from more than $9 per hundredweight in 1882 to $1 in 1887 (Wallace, 1936).

The boom-and-bust period for the beef industry coincided with an increase in the human population west of the Mississippi River. This introduced a transition for rangelands as the land was quickly converted to cropland. In the 30 years between 1870 and 1900, farmers brought more new land into cultivation—174 million hectares (430 million acres)—than had been brought into cultivation in the 250 years since the settlement of the Jamestown colony in Virginia (Athearn, 1986).

The Homestead Acts, which began in 1862, encouraged settlers to cross the Great Plains, taking with them farming methods better suited to eastern soils. Farmers plowed over natural short grasses to plant wheat and other grains and cereals. A few years of favorable weather and good yields bolstered enthusiasm for crop production, and farmers seemed to

continued

livestock. They also lacked the experience and knowledge needed to use properly the arid lands of the western United States. The number of livestock on U.S. rangelands expanded in the latter half of the nineteenth century, at the same time that many areas in the western United States suffered severe droughts. The combination of too many livestock, improper management practices, and drought accelerated the rate of soil erosion; depleted the amount of forage; and altered the species composition, density, and production of rangeland vegetation over extensive areas of the western United States.

Federal Management of U.S. Rangelands

This early crisis on U.S. rangelands led to efforts to bring large areas of western rangelands under the jurisdiction of the federal government, beginning with the creation of the forest reserves in 1891 and the USFS in

---

pay little attention to the fact that, in many areas, the fragile topsoil was shallow and was threatened by wind and water erosion (Athearn, 1986).

In 1879, John Wesley Powell wrote the "Report on the Lands of the Arid Region of the United States." He warned that low levels of rainfall made traditional large-scale farming impractical beyond the hundredth meridian, which roughly divides the nation through the Dakotas, Nebraska, Kansas, Oklahoma, and Texas. He proposed two alternatives: small irrigated farms or large grazing farms with small sections that could be irrigated (Powell, 1879). But his report was largely ignored by policymakers who did not understand the arid western landscape.

In California, more farmers turned from cattle to crops, and by 1889, California was second only to Minnesota in the production of wheat (Paul, 1988). Planting of as much as 4,000 hectares (10,000 acres) per farm became more common as the work was eased by combines hauled by steam-powered tractors. Low property values and minimal taxes on unimproved land encouraged farmers to plow and plant.

Finally, by the early 1930s the Great Plains had suffered through a decade of drought and people began to realize that they needed to manage and conserve the land better. This was the third period of human interaction on the rangeland, which was a time of reassessment of the land and its resources and debates over its use.

In 1931, during a national conference on land use in Chicago, Secretary of Agriculture Arthur M. Hyde spoke of the need to use better land management practices. Three years later, the Taylor Grazing Act established the federal administration of about 32 million hectares (80 million acres) of rangeland. President Franklin Roosevelt's New Deal programs, such

1905 and culminating in the Taylor Grazing Act of 1934. The Grazing Service, later to become BLM, was created in 1934 to manage these new federal lands, and the Soil Conservation Service (SCS) of USDA was created in 1935 to provide technical range management assistance to private landowners.

Changing perceptions of which values of rangeland ecosystems are most important have stimulated new debates over whether these public lands should be used to produce livestock, to support wildlife, to improve water quality, or for recreational purposes and how much of each of these uses was appropriate. A wave of environmental legislation—including the Multiple Use and Sustained Yield Act in 1960, the Resources Planning Act of 1974, the Federal Land Policy and Management Act of 1976, and the Soil and Water Resources Conservation Act of 1977—was enacted at least in part in response to concern about the state of U.S. rangelands.

---

as the Civilian Conservation Crops, brought more-sophisticated large-scale water management and irrigation practices to the western United States.

The Great Plains Drought Committee was formed in 1936, the same year that the secretary of USDA wrote to the Senate, highlighting the need to revitalize the rangelands while acknowledging changing demands for the land, including watershed and wildlife protection and the provision of recreational space.

The national parks were established, although they, too, frequently inspired debate, as with the creation of the Jackson Hole National Monument in Wyoming. The federal government wanted to buy up land for the monument, but local landowners complained that the project would unfairly deprive them of rangeland. As a compromise, the government reduced the size of the proposed project but bought more land to add to nearby Grand Teton National Park (Athearn, 1986).

During the post-World War II year, the tourism industry flourished in the western United States because of the favorable, dry climate and scenic attractions. Through the persistence of conservationists, who fought to keep much of the western land for public use, President Lyndon B. Johnson signed the Wilderness Act of 1964. Since passage of the Wilderness Act, millions of hectares have been set aside as wilderness areas. Today, many of those lands, as well as most other federal rangelands, are used for a variety of recreational enterprises, such as hiking, camping, horseback riding, and skiing, and they are still grazed by a restricted number of livestock.

Present State of Rangelands

Although most observers agree that rangeland degradation was widespread on overstocked and drought-plagued rangelands at the turn of the century, the current conditions on U.S. rangelands are a matter of sharp debate.

ASSESSMENTS OF RANGELANDS

Some reports have concluded that widespread historical degradation of rangelands has been halted and that rangelands, for the most part, have been recovering in the latter half of this century. For example, Box (1990) applied his professional judgment to data on trend (a change in a certain characteristic of rangeland over time) in range condition (SCS) and ecological status (USFS and BLM) ratings; he concluded that widespread degradation had been halted by the 1930s and that the trend in range condition (SCS) has generally been upward since that time. (See Chapter 3 for discussions of range condition and ecological status.)

In its most recent report on the state of the public rangelands, BLM (U.S. Department of the Interior, Bureau of Land Management, 1990) echoed the conclusions of Box (1990). It also reported that the current trend is stable or improving on more than 87 percent of public rangelands.

The Society for Range Management (1989) reviewed data provided by BLM, USFS, and SCS. It reported an improvement on 15 percent, a decline on 14 percent, and no apparent trend on 64 percent of the lands administered by BLM. Comparable data for lands administered by USFS were 43, 14, and 43 percent, respectively. The respective values for nonfederal rangelands were 16, 14, and 70 percent.

Other reports have described the continuing problems of rangeland degradation. For example, the National Resources Inventory, which is conducted once every 5 years by SCS, reported that in 1987 about 20 million hectares (49 million acres) of nonfederal rangelands (12 percent) were eroding at greater than the soil loss tolerance level and that over 11 million hectares (27 million acres) were eroding at twice the soil loss tolerance level (U.S. Department of Agriculture, Soil Conservation Service, 1989b) (Figure 1-2). (The soil loss tolerance level is the estimated maximum annual rate of erosion that can be tolerated without damaging soil productivity.) These data included rangelands eroding because of water-caused sheet and rill erosion only. (Sheet erosion is erosion caused by water running off unprotected soil in thin sheets; and rill erosion is that caused by water running off unprotected soil in small channels called rills.) Other forms of water erosion, such as gullying, combined with wind erosion, undoubtedly damage millions of acres of rangelands as well. No comparable data are available for federal rangelands, but there is no reason to assume that erosion is less severe on federal lands.

Environmental impact statements prepared by BLM and USFS reported rangeland degradation from soil erosion (U.S. Department of Agriculture, U.S. Forest Service, 1990b; U.S. Department of the Interior, Bureau of Land Management, 1987b), soil compaction (U.S. Department of the Interior, Bureau of Land Management, 1987c), the spread of introduced weed species (U.S. Department of the Interior, Bureau of Land Management, 1985b), reduced water quality and wildlife habitat (U.S. Department of Agriculture, U.S. Forest Service, 1991a; U.S. Department of the Interior, Bureau of Land Management, 1983a,b), and degradation of riparian habitat (U.S. Department of the Interior, Bureau of Land Management, 1983a, 1985c).

The U.S. General Accounting Office (GAO) issued two reports in 1988. The first one (U.S. General Accounting Office, 1988b) was based on a survey of BLM and USFS range managers. It reported that 19 percent of the BLM and USFS grazing allotments may be threatened with further degradation because of overstocking and that the condition of 8 percent of the grazing allotments was actually declining. The second report (U.S. General Accounting Office, 1988a), which was based on available data for riparian areas, reported that although some riparian areas were successfully restored, many thousands of kilometers of riparian habitat were in

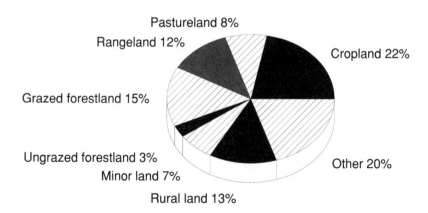

Pastureland 8%
Rangeland 12%
Cropland 22%
Grazed forestland 15%
Ungrazed forestland 3%
Minor land 7%
Other 20%
Rural land 13%

FIGURE 1-2 Categorized by land use, the chart shows 1987 figures for percentage of land eroding, by sheet and rill erosion alone, at higher than soil loss tolerance levels. Source: Adapted from U.S. Department of Agriculture, Soil Conservation Service. 1989. Summary Report: 1987, National Resources Inventory Statistical Bulletin No. 790. Washington, D.C.: U.S. Department of Agriculture.

need of treatment. Chaney and colleagues (1990), in a report prepared for the U.S. Environmental Protection Agency reported that "extensive field observations in the late 1980s suggest riparian areas throughout much of the west were in the worst condition in history" (Chaney et al., 1990:5).

DEFICIENCIES OF ASSESSMENTS

All of these reports, regardless of their conclusions, have been criticized by various interests, and none are based on comprehensive inventories of rangelands. The Society for Range Management (1989), for example, cautioned that the data available for their analysis were collected in the 1960s and 1970s and that the agencies did not have current data to support their professional opinion that the rangelands under their jurisdiction has improved during the past 20 years. Similarly, Box (1990) cautioned that trend data were not available for 12 percent of the national forests and 26 percent of rangelands managed by BLM. The utility of the models used to estimate erosion on rangelands in the National Resources Inventory has been questioned (U.S. Department of Agriculture, Soil Conservation Service, 1989a), data from environmental impact statements do not represent all rangelands, and the results reported by GAO are based on surveys of professional opinion rather than surveys of rangelands.

All national assessments suffer from the lack of current, comprehensive, and statistically representative data obtained in the field. No data collected using the same methods over time or using a sampling design that enables aggregation of the data at the national level are available for assessing both federal and nonfederal rangelands. Many reports depend on the opinion and judgment of both field personnel and authors rather than on current data. The reports cited above attempted to combine these data into a national-level assessment of rangelands, but the results have been inconclusive.

UTILITY OF CURRENT METHODS AND DATA

The debate is further clouded by disagreement within the scientific community over the utility of the current range condition (SCS), ecological status (USFS and BLM), and apparent trend assessment methods (see Chapter 3). The methods developed in the early 1900s were designed to assess the suitability of rangelands for grazing. New methods were adopted after 1950, but some ecologists now challenge the validity of those methods for assessing rangelands. Even where representative surveys of rangelands have been conducted using current range condition (SCS) or ecological status (USFS and BLM) ratings, such as in the National Resources Inventory, the utility of the results as measures of the status of rangelands is now in question.

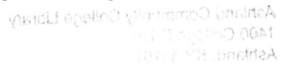

This disagreement and uncertainty concerning the state of the U.S. rangelands have become inextricably bound to the debate over the proper use and management of federal rangelands administered by BLM, USFS, and other agencies of the federal government. Public concern over the effect of livestock grazing on federal rangelands has intensified (Royte, 1990; Shaw, 1990; Wuerthner, 1991), leading to a variety of current efforts to restrict livestock grazing on federal lands.

The fact that available data do not allow investigators to reach definitive conclusions on the relative proportions of rangelands that are improving or degrading or on the relative rates of improvement or degradation seriously impedes efforts to resolve the debate over proper use and management of the nation's federal and nonfederal rangelands. The data that have been available for assessing the status of rangelands are obtained by many different methods and from many different sources. Different experts who look at the same data have interpreted them differently, confusing both the public and rangeland professionals.

URGENT NEED FOR NATIONAL ASSESSMENTS

There is an urgent need to develop the methods and data collection systems needed to determine whether rangelands are improving or degrading. The importance of the commodities and values provided by

Many wild animals graze on U.S. rangelands. Here, elk graze in a California state park. Credit: USDA U.S. Forest Service.

rangelands, the history of rangeland degradation, evidence of current degradation, and the inadequate data on current rangeland conditions suggest that it is unwise to neglect the status of the nation's rangelands. All attempts at national-level assessments reveal that degradation, particularly that from wind and water erosion, occurs on a significant portion of the nation's rangelands. The area of rangelands estimated to be deteriorating varies depending on the data that are used and how they are interpreted. The fact that it is impossible, with current methods and with current data, to determine whether federal and nonfederal rangelands are improving or degrading is itself a cause for concern.

Given the ecological and economic importance of U.S. rangelands, it is important that their capacity to satisfy values and produce commodities be conserved. Overgrazing by domestic or wild animals, inappropriate recreational uses, disease and insect outbreaks, drought, and other human-induced or naturally occurring stresses can and do degrade rangelands. Serious degradation can result in the irreversible loss of the capacity of rangelands to produce commodities and satisfy values and the loss of some or all options for using and managing rangelands in the future.

Federal and nonfederal rangelands produce a diversity of tangible commodities and satisfy many societal values that are important to the U.S. economy and the well-being of U.S. citizens. Overgrazing, harmful recreational uses, drought, and other human-induced or natural events have led to serious rangeland degradation in the past. Although the available data show that some rangelands continue to deteriorate, the full extent and the causes of that degradation are the subjects of debate. Given the importance of rangelands and the potential for serious degradation from both mismanagement and natural events, it is essential that the responsible agencies marshal the resources needed to develop and implement the data collection systems needed to provide policymakers, ranchers, environmentalists, and the general public with more definitive information on the state of federal and nonfederal rangelands.

2 Rangeland Health

The choice of methods and criteria to assess rangelands depends on the questions the assessments are intended to answer. There are many different questions that assessments could be and have been designed to answer including: What is the quality and quantity of the livestock forage produced? Is habitat for wildlife improving or degrading? and What quantity and quality of water can rangeland watersheds be expected to provide? These are all important questions and, for the most part, answers to each of them require different information about rangelands.

Although essential, defining the purpose of national assessments of rangelands is not simply a scientific problem because the motivations for national assessments emanate from the reasons society values rangeland ecosystems. Institutionalizing goals for rangeland assessments, then, unavoidably entails a judgment about what information about rangelands is most important to provide national policymakers, ranchers, environmentalists, and the general public. The capacity of rangelands to satisfy the values of and produce commodities for these diverse groups depends on the integrity of the soil and ecological processes of rangelands. National assessments should provide accurate and accessible information about the status of rangeland ecosystems to all individuals and groups who have an interest in the values and commodities that rangelands provide. Providing this information should be the first step toward formulating decisions about the management and use of federal and nonfederal rangelands.

GOALS FOR NATIONAL ASSESSMENTS

Chapter 1 described the diverse values and commodities that rangelands provide. The capacity of rangelands to sustainably produce com-

Arrowleaf balsamroot (*Balsamorhiza sagittata*)

modities and satisfy values depends on the interaction of climate, plants, and animals in a particular geological and topographic setting over time. These interactions result in soil development and the production of particular kinds and amounts of vegetation and enable rangelands to adjust to changes in their environment or management. These interactions also give rangelands the ability to resist the destructive effects of such extreme events as droughts and intense rainstorms.

In other agricultural ecosystems, such as intensively managed croplands, the capacity to produce resources and satisfy values is often augmented by using high levels of external inputs such as irrigation water or fertilizer, the physical environment is modified by tillage or terracing of the land, and pests are controlled by applying chemical pesticides. Rangelands, for the most part, do not receive such inputs. The capacity of rangelands to produce commodities and satisfy values depends on the integrity of nutrient cycles, energy flows, plant community dynamics, an intact soil profile, and stores of nutrients and water.

Overgrazing, harmful recreational activities, disease and insect outbreaks, drought, and other factors can degrade rangeland health and, hence, the quantity and quality of the values and commodities that are provided. Rangeland degradation can result in an irreversible loss of the capacity to produce commodities and satisfy values and the loss of future options to use and manage rangelands.

The importance of protecting and sustaining the capacity of rangeland ecosystems to provide the values and commodities desired by society has been repeatedly recognized in national legislation. (See Chapter 5 for a more complete discussion of national legislation pertaining to rangeland assessments.) The Soil Conservation Service (SCS), the U.S. Forest Service (USFS), and the Bureau of Land Management (BLM) have been mandated to provide the assessments of rangeland ecosystems needed to protect the quality and sustained yield of renewable resources. The Environmental Protection Agency is developing the Environmental Monitoring and Assessment Program (EMAP) to monitor changes in U.S. ecosystems, including rangelands. Providing policymakers and the public with the information needed to determine whether the capacity of rangelands to satisfy values and produce commodities is being sustained, improved, or degraded should be the primary goal of national assessments of rangelands.

STANDARDS FOR RANGELAND ASSESSMENTS

The long-running debate over the use and management of rangelands has intensified recently. The debate largely centers on whether grazing is an appropriate use of federal rangelands, whether grazing is being properly managed by ranchers and the agencies responsible for managing federal rangelands (BLM and USFS), and whether grazing is degrading federal and nonfederal rangelands (Royte, 1990; Shaw, 1990; Wuerthner, 1991). At the same time that public debate over the use of the nation's rangelands has grown, a scientific debate over the use of current range condition (SCS) and ecological status (USFS and BLM) ratings as broad assessments of the ecological condition of rangelands has emerged and intensified. As recently as 1989, range condition (SCS) and ecological status (USFS and BLM) ratings have been interpreted as measures of rangeland health (U.S. Department of Agriculture, Soil Conservation Service, 1989a; Society for Range Management, 1989). Now, however, the scientific debate over the utility of current range condition (SCS) and ecological status (USFS and BLM) ratings has intensified, leading to disagreements over the proper interpretation of past and ongoing range condition (SCS) and ecological status (USFS and BLM) ratings.

Pendleton (1989), for example, has asserted that there is a direct connection between the assessment methods currently used by SCS (referred to as range condition classes) and many important characteristics of rangeland ecosystems.

> There is a direct relationship between secondary succession [succession occurring on land where the original vegetation was disturbed], range condition [as determined by current methods], and the conservation of soil, water, and related range resources. Although the relationships are not exact and precise, it can reasonably be inferred that depleted ranges, those in the lower condition classes, are producing less forage than they are capable of, that erosion is higher than is normal or proper, that wildlife habitat values are less than optimal, and that range hydrology is impaired. (Pendleton, 1989:30-31)

In contrast, Smith (1989) questioned whether there is any connection between range condition classes and rangeland health.

> Current methods do not distinguish between areas where site deterioration is occurring and those where it is not. . . . The general public assumes that (1) ranges in poor to fair condition [classes] got that way due to overgrazing by livestock, (2) these ranges are deteriorating, and (3) reduction or removal of livestock will improve the range condition (Comptroller General, 1977; Sharpe, 1979; U.S. Department of the Interior, 1979). Economists often assume that range condition [range classes] is related to livestock production or wildlife values and that the greatest return will come from improving poor condition ranges (Martin, 1984). All these assumptions are logical, but incorrect. (Smith, 1989:125)

Such divergent views on the proper interpretation of current rangeland classification and inventory methods have confused the debate over

Rangelands in Transition: The Jornada Experimental Range

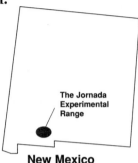

New Mexico

STARTING POINT: Black grama grassland
TRANSITION DYNAMICS: • Reduced grass cover
                                   • Soil compaction
                                   • Erosion and runoff
                                   • Increased patchiness of soil, water,
                                     and nutrients
END POINT: Desert shrubland

The shift from black grama (*Bouteloua eriopoda*) grassland to desert shrubland that occurred on large portions of the Jornada Experimental Range illustrates how a rangeland can cross a threshold because of the interaction of several factors. At this site, heavy grazing and moderate drought combine to reduce grass cover, compact the soil, increase erosion and runoff, and create a patchy distribution of water and nutrients that leads to an increased density of mesquite and creosote bushes. Once the shift is made to a patchy distribution of mesquite and creosote bushes, the ecosystem does not revert to the original black grama grassland without human intervention.

The 78,266-hectare (194,000-acre) Jornada Experimental Range is part of the Chihuahuan Desert, which extends from the south-central United States into Mexico. The Jornada Experimental Range has a mean annual temperature of 15.6°C (60.1°F) and a mean annual rainfall of 21 centimeters (about 8 inches). It is in an area of frequent floods and dust storms.

Black grama grassland at the Jornada Experimental Range is characterized by excellent soil coverage, which offers excellent protection from erosion. Black grama grass has a shallow root network that takes most of its water from the topsoil, minimizing the amount of moisture that can reach the subsoil. The uniform ground cover of grasses creates a uniform distribution of moisture and nutrients throughout the soil profile.

During periods of moderate drought, which may be difficult for the grassland to withstand, the introduction or intensification of grazing can decrease the black grama grass ground cover. Bare patches, which may have developed during the drought, become worse. Topsoil is compacted by trampling, less rainfall infiltrates the topsoil, and erosion accelerates. At this point, the rangeland is at risk of shifting to desert shrubland. If grazing pressure is reduced or the drought breaks, however, black grama would be expected to cover the bare patches without any further human intervention and the transition to desert shrubland would be prevented.

With the continuation of overgrazing during the drought, however, fundamental changes in the distribution of soil nutrients and moisture occur. These changes affect the soil and vegetation of the rangeland. Erosion becomes a significant problem as rain and nutrients are carried from the spreading bare spots and are deposited in depressions on the landscape. Creosote and mesquite shrubs begin to appear in the depressions where soil and nutrient deposits accumulate throughout the soil profile at a greater than average density. Pedestaling occurs around established, deeply rooted plants.

These changes are self-reinforcing. Soil around the grasses loses more nutrients and moisture, restricting the regeneration of grasses. Bare patches of soil release their available ammonia into the atmosphere, and denitrification of the soil intensifies after rainfalls. Runoff deposits nutrients to nurture young bushes and trees. Significant fire cannot be sustained on the patchy bare areas because of the lack of flammable organic material. Without natural burning, saplings and shrubs mature and proliferate. The black grama grassland has crossed a threshold to a desert shrubland. Even with human intervention, it may not be possible for the land to return to a black grama grassland, since soil nutrients and moisture essential to the grasses are limited to isolated pockets in which shrubs are firmly established.

At the Jornada Experimental Range, a range scientist inspects Lehmann lovegrass. Researchers removed invading brush and shrubs, replacing them with native and forage grasses. Credit: USDA Agricultural Research Service.

proper rangeland management. An agreed-to standard that can be used to determine whether the capacity of these rangelands to satisfy values and produce commodities is being conserved, degraded, or improved is needed. The lack of a consistently defined standard for acceptable conditions of rangeland ecosystems is the most significant limitation to current efforts to assess rangelands. The lack of such agreed-to standards has and continues to confuse the public, the U.S. Congress, ranchers, and range scientists themselves.

Rangeland Health

Rangeland health should be defined as the degree to which the integrity of the soil and ecological processes of rangeland ecosystems are sustained.

The capacity of rangelands to produce commodities and satisfy societal values depends on the interactions of climate, plants, and animals in a given physical landscape over time. These interactions are mediated by the soil and by internal ecological processes such as nutrient cycles, energy flows, and plant community dynamics. The integrity of the soil and ecological processes determines the vegetation, habitat, aesthetics, and other commodities and values that rangelands can provide and determines how well rangelands are able to resist the destructive effects of mismanagement or natural disturbances.

Rangelands are ecosystems not individual organisms and the use of the term "health" should not imply that simple analogies can be made between the health of an organism and the health of an ecosystem. Health, however, has been used to indicate the proper functioning of complex systems and is increasingly applied to ecosystems to indicate a condition in which ecological processes are functioning properly to maintain the structure, organization, and activity of the system over time. Recently, for example, Haskell et al. (1993) defined ecosystem health in terms of sustainability and stability, suggesting that an ecological system should be considered healthy if the system is "active and maintains its organization and autonomy over time and is resilient to stress" (Haskell et al., 1993:9).

The concept of forest health has frequently been used to refer to the effect of pests, pathogens and toxic compounds on the growth, development, and reproduction of forest communities (Brooks, 1992; Johnson et al., 1992). Use of the term "forest health," however, increasingly refers to a broader conception of ecosystem health that recognizes the importance of changes in nutrient cycles, soil attributes, air quality, and other structural and functional characteristics of forest ecosystems (Burkman and Hertel, 1992; Commission of the European Communities, Directorate General for Agriculture, 1990; Gray and Clark, 1992). Similarly, the concept of crop health is increasingly being expanded to include the health of the agroecosystem as a whole (Cook and Veseth, 1991). Efforts to develop measurable indicators of change in ecosystems as part of ecological risk assessments has also increased the use and development of the concept of ecological health as a measure of the integrity of the structure and function of ecosystems (International Joint Commission, 1991; National Research Council of Canada, 1985; Schaeffer et al., 1988).

Webster's Third New International Dictionary defines *healthy* as "(1) functioning properly or normally in its vital functions, (2) free from malfunctioning of any kind, and (3) productive of good of any kind." The terms "healthy" or "unhealthy" are most properly applied to ecosystems as an indication of proper or normal functioning of ecological processes resulting in the production of good, that is commodities or values, that are important to private landowners and the public at large.

The term "health," then, as used by the committee, is an indication of the ecological integrity of rangeland ecosystems. The term "ecological integrity" has recently been defined as "maintenance of the structure and functional attributes characteristic of a particular locale, including normal variability" (National Research Council, 1992:520). More specifically, the committee recommends the term "rangeland health" be used to indicate the degree of integrity of the soil and ecological processes of rangeland ecosystems that are most important in sustaining the capacity of rangelands to satisfy values and produce the commodities.

Determining whether the capacity of a rangeland to satisfy values and produce commodities is being sustained will not resolve the debate over the proper use and management of that rangeland. A separate system is needed to evaluate use of a particular rangeland and the kind and amounts of vegetation needed to support that use (Ellison, 1949; Friedel, 1991; Humphrey, 1947; Lauenroth, 1985; Laycock, 1989; Shiflet, 1973; Society for Range Management, Range Inventory Standardization Committee, 1983; Tueller, 1973; West, 1985; Wilson, 1989). If, however, the public, policymakers, ranchers, and range managers can be assured that rangeland health is conserved, the debate can profitably shift to whether rangelands are best used for the production of livestock, wildlife, or recreation or some combination of these. These decisions will be contentious, but they can at least be made in the context of conserving the health, and therefore the capacity, of rangelands to produce commodities and satisfy values, regardless of their use.

Categories for Rangeland Assessments

Rangeland ecosystems are dynamic systems, and fitting rangelands into categories based on ecological criteria is a difficult but essential task for national assessments of rangelands. The capacity of rangelands to produce commodities and satisfy values depends on the integrity of soils and ecological processes, that is, on their health. Range managers, policy-makers, and the public need to know whether the health of federal and nonfederal rangelands is being sustained, improved, or degraded. This requires defining boundaries between states of rangelands depending on the degree to which the integrity of the soil and internal ecological processes are protected.

The principal purpose of the rangeland inventories completed by SCS, BLM, and USFS should be to determine the proportion and location of rangelands that are healthy, at risk, or unhealthy.

The categories defined for purposes of national rangeland assessments should facilitate the interpretation of the results of those assessments for policymakers, range managers, ranchers, and the public. The categories used for national assessments should signal where rangeland management or technical assistance are needed to prevent degradation or to improve damaged rangelands.

The committee recommends that rangelands be placed in three broad categories based on an evaluation of ecological health. Rangelands should be considered (1) healthy if an evaluation of the soil and ecological processes indicates that the capacity to satisfy values and produce commodities is being sustained, (2) at risk if the assessment indicates an increased vulnerability to degradation, and (3) unhealthy if the assessment indicates that degradation has resulted in an irreversible loss of capacity to provide values and commodities.

DEFINING BOUNDARIES

Categorizing rangelands as healthy, at risk, or unhealthy requires defining two boundaries: the boundary distinguishing healthy from at-risk rangelands and the boundary distinguishing at-risk from unhealthy rangelands. Rangelands, however, are constantly adapting in response to changes in physical environment, use, and management and to episodic events such as fires, droughts, and intense rainstorms. These constant adaptations are reflected in changes in many characteristics of the rangeland ecosystem such as plant composition, the amount of plant biomass produced, the amount of nutrients and the rate at which they are cycled, and the amount and composition of soil organic matter. The ecological state of a rangeland at a given point in time is the sum of these characteristics. The rangeland ecosystem shifts between different ecological states over time in response to natural or human-induced factors. Such changes can be sudden or they may occur gradually.

There are important differences between processes of change, however, that can be used to identify boundaries between healthy, at-risk, and unhealthy rangelands for the purposes of national assessments. Some changes in ecological state may have no long-term effect on the capacity of the rangeland to produce commodities or satisfy values. A change in the relative abundance of the dominant plant species, for example, may reflect seasonal variation in rainfall rather than a change in the capacity of the rangeland to produce wildlife habitat or forage. Other changes can be destructive, but their destructive effects can be reversed by changes in use and management or as natural conditions improve if the integrity of the soil and ecological processes has been conserved. Still other changes—soil degradation, the interruption of nutrient cycles, and the loss of important species or seed sources, for example—can lead to irreversible changes that reduce the amount and diversity of vegetation, habitat, aesthetics, and other commodities and values. Once these changes have occurred, external inputs, reseeding, or soil reclamation, for example, will be required to restore the rangeland to a healthy state. Even with restoration, however, some loss of capacity to produce commodities and satisfy values may be permanent.

The boundaries between healthy, at-risk, and unhealthy states of a rangeland should be distinguished based on changes in the soil and ecological processes that determine (1) the capacity of the rangeland to produce commodities and satisfy values and (2) the reversibility of the changes between states and can be illustrated by a model (Figure 2-1). The boundary between at-risk and unhealthy states should indicate a reduction in capacity to satisfy values and produce commodities that is difficult to reverse without substantial external inputs. The boundary between healthy and at-risk states should indicate a reduction in capacity to provide values and commodities that is likely to be reversed through changes in use and management or as natural conditions improve. These boundaries are referred to in Figure 2-1 as the *threshold of rangeland health* and the *early warning line*. A brief discussion of the processes leading to changes in ecological state will help clarify the distinctions between healthy, at-risk, and unhealthy rangelands.

THRESHOLDS BETWEEN ECOLOGICAL STATES

A threshold can be defined as a boundary in space and time between two ecological states. Ecologists have recognized and studied how ecosystems change from one state to another across thresholds (Holling, 1973; Wissel, 1984). Threshold changes involve shifts in plant composition; changes in the physical, chemical, or biological properties of soils; or changes in basic ecological processes such as nutrient cycles. They are different from other changes from one state to another because they are

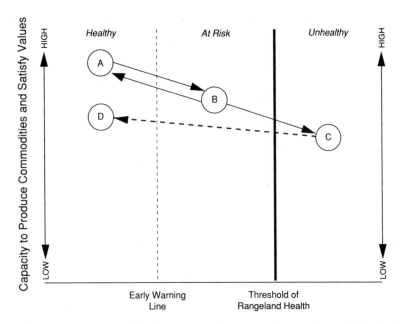

FIGURE 2-1   A simple model of transitions along a continuum of rangeland health.

not reversible on a practical time scale without human intervention. In some cases, human intervention may not be sufficient to reverse these changes (Friedel et al., 1990), for example, severe soil erosion.

Interaction between factors often accelerates changes in rangelands. Changes in grazing management can result in rapid positive changes in the composition of plant communities and the amount of annual biomass that is produced if the changes in use and management are accompanied by a series of years with above-average precipitation. Changes in the vegetation will occur more slowly or may not occur at all under climatic conditions that are not favorable for seedling establishment and growth. Similarly, overgrazing that coincides with drought years can result in the rapid degradation of rangeland health. A high-intensity rainstorm that occurs at a time when plant cover has been reduced by overgrazing, fires, or droughts can result in accelerated rates of soil erosion.

PROCESSES OF CHANGE

Ellison (1949) distinguished two types of rangeland change—secondary succession and destructive change. (Primary succession is a series of changes in the composition of the plant and animal life of a particular area. Secondary succession occurs in places where the original vegetation has been disturbed, for example, on land affected by fire or drought.) According to Ellison, secondary succession entailed changes in plant composition, such as a decline in the number of plant species that were palatable to grazing animals or an increase in the shrub component because of grazing pressure. These changes in plant composition were considered normal adjustments as a result of grazing.

Ellison considered destructive change to be beyond the limits of normal change and to be induced by accelerated erosion, which was evidence of a basic change in the relationship between components of the rangeland ecosystem—a change of drastic proportions over and above the normal range of environmental stresses. Furthermore, once such change was initiated, it could not easily be reversed, even with the discontinuation of grazing. Destructive change represented a new process of change that is not comparable to the process of soil development and that results in the permanent loss of productive capacity. The process of destructive change, described by Ellison can be thought of as leading to a threshold shift between two ecological states. In this case, soil degradation leads to a reduction in the productive capacity of the rangeland that is difficult or impossible to reverse.

Friedel (1991) suggests that rangeland plant communities change in response to various combinations of different factors such as grazing season, the animal species present on the rangeland site, and variables related to the specific site. These changes do not preclude shifts to other short-lived plant communities if the grazing season and grazing pressures change. Various combinations of climatic and grazing conditions may, however, induce a change in plant species composition that is not readily reversible. In such cases, the system has crossed a threshold.

Friedel recognized two such changes in threshold—from grassland to woodland and from stable to degraded soil—on Australian rangelands. The first threshold change—from grass to woody vegetation—results when grazing reduces the density of the grass layer. Germinating woody plants displace grasses because their deeper roots are able to reach the water found in increasingly deeper subsoil layers. This change results in a transition across a threshold to woody vegetation that is difficult to reverse (Friedel, 1991).

The second threshold occurs when soil erosion irreversibly alters the physical, chemical, and biological properties of the soil. This results in reduced infiltration of rainfall into the soil. The land becomes too xeric (dry) for the establishment of grasses or woody plants. Well-established plants may remain and new plants may become established during infrequent periods when climatic conditions are particularly favorable (Friedel, 1991). Soil erosion thereby irreversibly changes the kind and amount of vegetation the site can produce.

Rangelands in Transition: South African Tall Grassveld

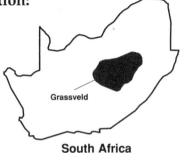

South Africa

STARTING POINT: Perennial grassland
TRANSITION DYNAMICS: • Reduced perennial grasses
                          • Increased erosion
                          • Reduced soil cover
END POINT: Annual grassland with bare areas

Tall grassvelds in South Africa can shift from a mix of palatable grasses to a mix of unpalatable perennial grasses and annual grasses, depending on grazing management (Westoby et al., 1989). This particular transition can be reversed with minimal human intervention. Prolonged overgrazing, however, in combination with factors such as increased erosion and reduced soil cover can cause a shift to a patchy mix of annual grasses and bare spots. Once this transition occurs, a return to the original state is likely to be difficult, if not impossible, without significant human intervention.

Light grazing decreases the cover of palatable perennial grasses, including *Themeda triandra* and *Eragrostis racemosa*, and increases the cover of unpalatable perennial or annual grasses, such as *Cymbopogon excavatus*. These changes signal that the rangeland is at risk of shifting across a threshold to a different combination of spe-

Shown are a spikelet and a tuft of Themeda triandra. *Credit: Photo by E. B. van Wyk, courtesy of the Embassy of South Africa.*

cies. A timely decrease in grazing pressure, however, allows a resurgence of the palatable grasses and a return to the original state.

If overgrazing continues, however, problems such as erosion and soil runoff combine to force the rangeland over a threshold. Palatable and unpalatable perennial grasses are reduced and bare patches appear. Annual grasses begin to replace perennial grasses. As erosion becomes more severe, the seed supply from perennial grasses is greatly reduced or even lost. The seeds that are still available cannot germinate because of

South African grassveld dominated by Themeda triandra. *Credit: Photo by E. B. van Wyk, courtesy of the Embassy of South Africa.*

insufficient nutrient-rich soil and lack of suitable seedbeds. These changes signal that the rangeland has crossed a threshold to a new ecological state. Simply reducing grazing pressure, once the threshold is crossed, is no longer sufficient to regain the initial state. Management techniques such as soil reclamation and reseeding with perennial grasses would be required, but they may not necessarily lead to a recovery to the initial state.

Episodic events can be significant causes of threshold changes. A period of above-average rainfall, for example, may facilitate the germination and growth of seedlings of woody plants. These woody plants may eventually dominate the rangeland unless fires occur before the seedlings become well established. Similarly, if rainfall is episodic, plants may have only infrequent opportunities to regenerate, and these opportunities may be decades apart. The season in which the rainfall occurs will influence which plant species produce seed and which seeds germinate and become established. The timing of an episodic rainfall event, then, may determine the plant composition of a rangeland for many years. Finally, a single storm producing large amounts of rain can also initiate the erosion of susceptible soils and alter the productivities of entire landscapes (Friedel et al., 1990).

Risser (1989) suggested that recovery of species composition following a disturbance may be slow either because the species that must increase have slow dispersal rates or because their seeds may germinate infrequently and their seedlings may have strict requirements for water, nutrients, or other factors that must be met if they are to grow. Succession may become suspended or static for long periods because of a lack of seeds or seed dispersal, dominance of a life-form that does not allow other species to increase or invade the area, specific physiological requirements that limit seedling establishment except, climatic changes, restriction of natural fires, or other factors (Laycock, 1989).

THRESHOLD OF RANGELAND HEALTH

The threshold of rangeland health should be defined as a boundary between ecological states of a rangeland ecosystem that, once crossed, is not easily reversible and results in the loss of capacity to produce commodities and satisfy values.

The threshold of rangeland health is distinguished from other boundaries between the ecological states of a rangeland ecosystem by two key factors. First, as the use of the term "threshold" suggests, the shift from one ecological state to another across the boundary is not easily reversed. Second, as the use of the term "health" suggests, changes in the soil or ecological processes result in a change in the capacity of the rangeland to satisfy values or produce commodities.

Significant external inputs, such as soil stabilization or reclamation, reseeding, or control of unwanted vegetation, are usually required for a rangeland to regain a healthy state once the threshold has been crossed. Simple management changes such as improved grazing or the reintroduction of fire will not restore rangeland health within a practical time frame when soil degradation, loss of seed sources, changes in vegetative structure of the plant community, disruption of nutrient cycles, or a combination of these and other factors are also involved.

Degradation of the soil and of ecological function, which leads to the transition from an at-risk to an unhealthy state, causes a reduction of the capacity to produce commodities and satisfy values. Given soil reclamation or reseeding efforts or other external inputs, transition across the threshold of rangeland health from unhealthy to healthy is possible. Even though health is restored, the rangeland may not produce the same mix and amount of resources and satisfy the same values as it did in the original healthy state.

EARLY WARNING LINE

The rangeland inventories and routine monitoring completed by SCS, BLM, and USFS should provide an early warning of rangelands that are vulnerable to a shift across the threshold of rangeland health.

An early warning of changes in soil or ecological processes that increase the vulnerability of rangelands to a shift across the threshold of rangeland health is essential to preventing loss of health. Degradation of soils or ecological function results in some loss of rangeland health and, therefore, the capacity to produce commodities and satisfy values as a rangeland changes from a healthy (state A) to an at-risk (state B) state (Figure 2-1). The transition from a healthy to an at-risk state is thought to be reversible, however, if the human-induced or natural factors that caused degradation are alleviated. A change from an at-risk to a healthy state, however, does not necessarily entail a return to the original plant community composition of the site.

The boundary between healthy and at risk can be thought of as an early warning and signals the need to take corrective action or further investigate the site. Identification of at-risk rangelands would enable range managers to take appropriate action before the health of the rangeland and the capacity of the rangeland to produce commodities and satisfy values is impaired.

The transition from a healthy rangeland (state A) to one at risk (state B) involves changes in the physical environment and biological systems that make the area more susceptible to near-permanent changes resulting from extreme climatic events, improper use or management, or other stresses. This risk may be due to an increased vulnerability to extreme events that cause a sudden transition across the threshold of health or to the cumulative effects of one or more ecological stresses. A rangeland with compacted soil or reduced plant cover, for example, may be vulnerable to rapid and irreversible gully formation during a torrential rainfall. Alternatively, a rangeland may approach the threshold of rangeland

Rangelands in Transition: The Rio Grande Plains

Texas

STARTING POINT: Perennial grassland
TRANSITION DYNAMICS: • Change in rainfall pattern
                             • Reduced grass cover
                             • Reduction in fire frequency
                             • Redistribution of soil moisture
END POINT: Woody shrubland

The historical change in vegetation on the Rio Grande Plains near Alice, Texas—from a savannah with only scattered trees to a subtropical thorny woodland—illustrates how several factors can interact to cause a rangeland to shift across a threshold. For example, a change in climate, such as a shift in rainfall pattern from frequent showers limited to small areas to infrequent storms across the rangeland; overgrazing, which reduces grass cover; a decrease in fire frequency; and changes in soil moisture can combine to have dramatic, far-reaching effects. Research indicates that the interactions of those factors are causing large-scale changes in vegetation in southern Texas (Archer, 1989).

The climate of the subject site at the Texas Agricultural Experimental Station in La Copita Research Area is subtropical, with mean annual rainfall of 68 centimeters (27 inches) and a mean annual temperature of 22.4°C (72.3°F). The soils are fine, sandy loams from sandstone. Initially classified as a *Prosopis-Acacia-Andropogon-Setaria* savannah, the site has been grazed by cattle since the late 1800s and has experienced a 23 percent increase in woody cover since 1941. At issue is the reason for the increase in the number of mesquite (*Prosopis glandulosa*) trees and the shrubs that surround them. Mesquite is becoming common and more dominant in ecosystems throughout the southwestern United States, although historically it was a minor component of rangeland ecosystems. This change to more mesquite trees and the corresponding shrub clusters continues, but the shift accelerated between 1960 and 1983.

Archer's (1989) research uncovered four major reasons for the change to a woodland: a change in rainfall patterns over the past 100 to 200 years, reduced grass cover, fewer fires, and a reduction in the available moisture in the topsoil. Any one factor alone would be insufficient to so dramatically alter the vegetation of the Rio Grande Plains; a combination of two or more of the suggested factors is probably required, according to Archer.

A shift from light and frequent to heavy and less frequent rainstorms or an increase in winter rainfall over that in summer increases the proportion of rain that infiltrates to deeper soil layers. This change in the ratio of topsoil to subsoil moisture is also caused by overgrazing, which reduces grass cover. Shallow-rooted grasses that capture most of the rainfall, keeping it in the topsoil, are reduced, allowing more rainfall to reach the subsoil. Mesquite, which can reach the subsoil moisture, therefore gains a competitive advantage over grasses, and individual shrubs become established. These changes signal that the rangeland is at risk of shifting to a woody shrubland if the trends continue.

At this point, reduced grazing pressure or fire may be sufficient to reverse the increase in shrubs. If no change in management or climate occurs, however, shrubs and trees become established in clusters around the mesquite trees. These clusters capture more and more of the available water and nutrients. The development of clustered woody vegetation reduces the chance of fire. Historically, fire sweeps across the grassland, burning seedlings in its path and permitting the regeneration of grasses, which grow more quickly after a fire. Because of the patchiness in the grassland, however, fire is extinguished in the bare spots because of the lack of flammable organic matter. Without fire, mesquite trees and their associated shrubs proliferate. Grazing contributes to this trend as cattle eat mesquite seeds, depositing them in dung, a nutrient-rich environment ideally suited to germination. The rangeland crosses a threshold, becoming a subtropical thorny woodland.

Grazing contributes to the proliferation of mesquite trees and their associated shrubs as cattle eat mesquite seeds, depositing them in dung, a nutrient-rich environment ideally suited to germination. Credit: USDA Agricultural Research Service.

health gradually as important seed sources are lost as a result of the combined effects of drought and poorly managed grazing. Lands considered at risk can become healthy once the climate has returned to a more average state or once use and management are appropriately changed.

MULTIPLE STATES AND TRANSITIONS

The model shown in Figure 2-1 suggests that pathways of loss and recovery of rangeland health are linear and simple. This is not necessarily the case. States A, B, C, and D may represent a complex of related plant communities rather than a single stable community (Figure 2-2). Change from one plant community to another within the complex may be caused by predictable successional processes or by climatic variability, insect and disease outbreaks, the grazing system used, or other factors. The plant communities within a complex may not produce the same mix of commodities or satisfy the same values. A short-lived drought or temporary heavy grazing, for example, may result in a reduction in biomass production. Increased amounts of rainfall or improved grazing management may produce a shift to a different plant community within a complex that produces a different mix of commodities and values. Changes within a

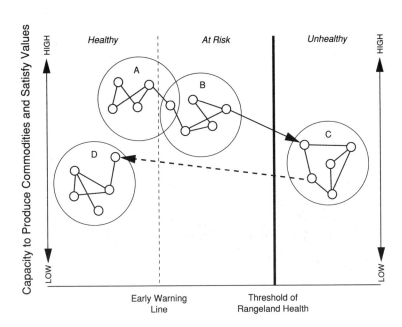

FIGURE 2-2 Expanded model of transitions along a continuum of rangeland health.

complex are reversible, as are transitions between complexes in states B and A.

In Figure 2-2, transition from state B to state C indicates an irreversible change that results in loss of capacity to produce commodities and satisfy values. There may be shifts from one plant community to another within complex C, but the transition from state C to state D will not occur without human intervention.

Transition from state B to state C in Figure 2-2 entails the loss of health. Continued soil degradation, disruption of nutrient cycles and energy flow, or the loss of species that are important functional components of the complex results in a transition that is difficult to reverse. The composition of the rangeland may continue to change within the new complex, but reversal of the degradation will require external inputs, such as soil reclamation and reseeding of vegetation. Even with such corrective action there may be permanent loss of capacity to produce commodities and satisfy values, at least within practical time frames and costs. Future options to use and manage the site may be lost as well.

The elements of the simple (Figure 2-1) or the expanded (Figure 2-2) model are the same. Both incorporate the idea of a transition across a threshold of rangeland health that is not easily reversible and that entails the permanent loss of capacity to produce commodities and satisfy values, even if corrective action is taken.

ROLE OF RANGELAND HEALTH IN RANGELAND MANAGEMENT

Although the concept of changes across ecological thresholds has long been recognized, the concept has not explicitly been included in assessments of rangelands. It is essential to understand the role that rangeland health assessments should play in the larger effort to classify, inventory, monitor, and manage the nation's rangelands. The concept of rangeland health should be only one part of a complete system for managing rangelands. No single index will meet the informational needs of managers, ranchers, policymakers, and the public.

Goal of Range Management

The minimum standard for rangeland management should be the prevention of human-induced loss of rangeland health.

Large investments of time, money, and energy are required to restore unhealthy rangelands. Even with restoration, there may be permanent loss of capacity to produce commodities and satisfy values or loss of options to use and manage those rangelands in the future. Any human-induced loss of rangeland health should be prevented.

The health of rangeland ecosystems may be affected by natural as well as human-induced factors. The ability of rangelands to produce commodities and satisfy values, for example, may be altered by long-term climate changes. Management of rangeland ecosystems must be sensitive to such changes, particularly changes that can threaten rangeland health, such as prolonged drought. So that human-induced loss of rangeland health is prevented, management and use of rangeland ecosystems will have to be adjusted if climate changes increase the vulnerability of rangelands.

Because rangelands are managed and used in ways that depend on the integrity of their soils and ecological processes, the fundamental aim of an assessment of the status of rangelands should be to determine whether a rangeland is healthy, at risk, or unhealthy. Rangelands found to be at risk should receive special attention. Management and monitoring of at-risk rangelands should be more intense than management and monitoring of healthy rangelands. Corrective action must be taken to prevent human-induced loss of rangeland health.

Additional Information Needed to Determine Appropriate Management

Rangeland health inventories and monitoring systems should be one part of a larger system of data gathering and analysis to inform range managers, policymakers, and the public.

No single method of evaluating rangelands will provide all the information needed by range managers, ranchers, policymakers, and the public. Rangeland health is a measure of the integrity of the soil and ecological processes. Other information will be needed to determine what causes the loss of health, what needs to be done to improve health, and how a particular rangeland should be used.

Rangeland health is a measure of whether the capacity of rangelands to produce commodities and satisfy values is being conserved. An assessment of rangeland health is not intended to quantify the suitability of particular rangelands for particular purposes. (The relationship between rangeland health and these uses is discussed in more detail in Chapter 3.) Quantification of the capacity of rangelands to produce specific commodities and satisfy particular values may require a separate assessment. The adoption of such an assessment system has been recommended by several experts and committees, most recently, the Society for Range Management's Task Group on Unity in Concepts and Terminology (1991).

Although rangeland health is related to the capacity of rangelands to produce commodities and satisfy values, the two concepts are different and are represented by different axes in Figures 2-1 and 2-2. One rangeland, because of its soil, climate, and topography, may be able to produce more total biomass annually than another rangeland. But both can be healthy. Differences between the capacities of different rangelands to produce commodities and satisfy values do not necessarily imply differences in health as defined here.

Rangeland health estimates the risk of the loss of the capacity to produce commodities and satisfy values by evaluating the integrity of the site's ecological processes and soils. Such an evaluation does not determine conclusively the processes that are responsible for the current state of health or determine what changes in management are required. Assessment of rangeland health may not be sufficient for a full understanding and treatment of many problems. It should serve as an early warning of problems and help the range manager decide whether detailed measurements are needed and, if so, what additional measurements need to be made.

Rangeland in Extreme Environments

There are landscapes where the prevailing environmental conditions constrain the development and conservation of the soil and ecological

The environmental conditions of South Dakota's Badlands have created a region characterized by intricate and sharp erosional sculptures, fantastically formed hills, labyrinthine drainage systems, and normally dry watercourses or arroyos formed of decomposed granite, loess, and other soft materials. Credit: Photo courtesy of the South Dakota Department of Tourism.

processes that indicate healthy conditions. Soils on these landscapes are unstable or absent, and nutrient cycles, energy flows, and other ecological processes are not established or are only poorly established. In such landscapes, a relatively stable biotic community has not gained a foothold and abiotic processes are more important than biotic processes. Rates of erosion are too great or rates of nutrient enrichment and organic matter accumulation are too slow to allow the development of soils. The Badlands of South Dakota and the Mancos-shale regions of Utah are examples. These landscapes are characterized by a lack of developed soils, by extreme climates, or both. The lack of developed soils may be due to the youth of these sites in geological terms, climates that are unfavorable to soil formation, or geological rates of erosion that preclude the development of soils.

Such sites would be considered unhealthy given the definitions proposed in this report. This unhealthy state, however, has not been induced by humans, and even the best management of these landscapes may not be sufficient to achieve healthy conditions. The primary concern of range management should be preventing loss of health on rangelands where human use and management can improve or degrade rangeland health. Even though sites such as the Badlands of South Dakota could be considered naturally unhealthy, they often satisfy important aesthetic and recreational values, and recreational use of these areas is often an important source of economic activity in the local area. Careful management may be required to protect their aesthetic and recreational values.

3 Current Methods of Rangeland Assessment

The current theory and practice of rangeland assessment have a long history that is closely related to the ways that rangelands were used and studied. The nineteenth century was a period of exploration and development of the rangelands of the western United States. The need for systematic methods of rangeland assessment first became apparent when Jared Smith was sent by the U.S. Botanical Survey in 1895 to study the causes of the deterioration of western rangelands that had been widely reported in the late 1880s. He reported that:

> The shortage of cattle all through the west is due to the fact that ranges were stocked up to the limit that they would carry during the series of exceptionally favorable years preceding the years of drought. Then followed the bad years when the native perennial grasses did not get rain enough to more than keep them alive. The cattle on the breeding grounds of the West and Southwest died by the thousands of thirst and starvation (Smith, 1896:322–323).

Such early investigations, however, were not based on a unifying science that could systematize the data collected to assess rangelands or relate the effects of livestock grazing to the rangeland deterioration that was evident in the late nineteenth century. The need for more thorough assessments was evident.

DEVELOPMENT OF CURRENT THEORY AND PRACTICE OF RANGELAND ASSESSMENTS

Between 1890 and 1905, 11 state agricultural experiment stations published 879 range management-related bulletins dealing with the control of weeds, pests, poisonous plants, soil moisture, fertility, conservation, rangeland inventory and analysis, water use, fencing, and other topics (Beetle, 1954). The U.S. Forest Service (USFS), which was formed in 1905,

Scarlet globemallow (*Sphaeralcea coccinea*)

51

Timeline of Rangeland Classification and Inventorying in the United States

Rangeland Use and Management	Year	Inventory
Spanish settlers introduce sheep and cattle into California rangelands.	1500	
British settlers expand livestock grazing to more rangelands in the western United States.	1800	
Major increase in livestock on rangelands in Texas, Kansas, Nebraska, Oklahoma, and the Great Plains.	1860	
Overgrazing of the Great Plains and the Great Basin is accompanied by drought and severe erosion.	1880	
	1895	Jared Smith conducts a survey of depleted western rangelands for the U.S. Botanical Survey. He reports that overgrazing and drought are responsible for widespread rangeland degradation.
		U.S. Department of Agriculture begins to publish reports on forage condition and grazing problems in the western United States.
U.S. forest reserves are established, bringing part of western rangelands under the jurisdiction of the U.S. Department of Agriculture.	1897	Surveys of forest reserves begin.
	1899	H. C. Cowles describes plant succession on the sand dunes of Lake Michigan.
	1900	State agricultural experiment stations begin to issue reports on the condition of rangelands and range management practices.

Rangeland Use and Management	Year	Inventory
U.S. Department of Agriculture establishes regulation of grazing on forest reserves.	1901	
The U.S. Forest Service (USFS) is formed and the National Forest system is created.	1905	National forests are surveyed.
	1905	F. E. Clements begins adapting succession to Great Plains grasslands.
	1910	A. W. Sampson begins the first ecologically based range work in the national forests of Utah.
	1917	J. T. Jardine develops the first scientific survey method.
	1923	A. W. Sampson introduces the use of succession as a way to assess grazing capacity.
Dust Bowl ravages western range-land and cropland.	1930	
	1933	A. R. Standing proposes quantifying rangeland assessments by measuring volume of vegetation rather than estimating plant cover.
The Taylor Grazing Act withdraws all remaining public land that is not under the jurisdiction of the USFS or other federal agencies into grazing districts under the jurisdiction of the Bureau of Land Management (BLM). BLM is charged with halting overgrazing and soil deterioration.	1934	
The Soil Conservation Service (SCS) is created. Its mandate is to inventory soil and water resources and to assist farmers and ranchers with ending erosion.	1935	

continued

Rangeland Use and Management	Year	Inventory
	1936	The U.S. Senate issues a report on the condition of rangelands and the causes of rangeland deterioration.
		The interagency range survey technique is standardized in the late 1930s and used by BLM, SCS, and USFS to inventory rangelands.
	1938	J. E. Weaver and F. E. Clements publish *Plant Ecology*, extending successional concepts explicitly to rangeland ecosystems. *Plant Ecology* becomes the standard text for the emerging field of range management.
	1949	E. J. Dyksterhuis solidifies the use of successional stages as measures of rangeland condition by proposing the measurement of range condition as a departure from climax vegetation for a specific range site. Dyksterhuis' range site and condition method becomes the basis for SCS rangeland assessments.
	1950	USFS develops the Parker three-step method of rangeland assessment.
		BLM continues to use the interagency survey method.
The Wilderness Act is passed. The National Forest Management Act mandates multiple use and sustained-yield policies for national forest management.	1964	
	1968	R. Daubenmire describes habitat types, which become the basis for USFS forest and rangeland classifications.

Rangeland Use and Management	Year	Inventory
National Environmental Protection Act requires all federal agencies to write environmental impact statements on major federal actions.	1969	
The Endangered Species Act requires federal agencies to protect listed wildlife species.	1973	
The *National Resources Defense Council* v. *Morton* requires environmental impact statements on all local grazing programs administered by BLM.	1974	
The Resources Planning Act requires USFS to inventory national forests every 10 years.		
Congress passes the Forest and Rangeland Renewable Resources Research Act, which provides the information needed for Resources Planning Act implementation.		
	1975	The first Resources Planning Act Assessment of resources, including rangelands on national forests, is published.
Congress passes the Federal Land Policy and Management Act, which requires BLM to prepare an inventory of the resources on federal lands under BLM's jurisdiction.	1976	
The Soil and Water Resources Conservation Act is passed, requiring SCS to inventory soil, water, wildlife habitat, and related resources on nonfederal lands. SCS establishes the National Resources Inventory to carry out its mandate.	1977	

continued

Rangeland Use and Management	Year	Inventory
The Public Rangeland Improvement Act calls for improvements in soil quality, wildlife habitat, watersheds, and vegetation on federal rangelands and requires inventories of federal rangelands.	1978	
	1979	BLM develops the soil-vegetation inventory method; the method is tested but is not formally adopted by BLM.
	1980	The second Resources Planning Act Assessment is published.
	1981	The first Resource Conservation Act Appraisal of soil and water resources on nonfederal lands based on the 1977 National Resources Inventory is published by SCS.
	1983	BLM adopts SCS range site and range condition procedures for assessing rangelands. The Society for Range Management recommends that SCS, BLM, and USFS adopt common terminology to classify rangelands and make ecological status ratings.
	1985	The National Resources Defense Council and the National Audubon Society assemble data on the rangelands under BLM's jurisdiction and report that many rangelands are in unsatisfactory condition.
	1987	The second Resources Conservation Act Appraisal of soil and water resources on nonfederal lands, based on the 1982 National Resources Inventory, is published by SCS.

Rangeland Use and Management	Year	Inventory
	1989	The Society for Range Management assembles data from SCS, BLM, and USFS in an attempt to make a national assessment of rangelands. The society reports that the data available for federal rangelands are not adequate for a national assessment. The USFS publishes an assessment of the range resources of the United States.
	1990	The third Resources Planning Act Assessment is published.

recognized the need to develop a scientifically credible and economically feasible method of surveying rangelands to carry out its mandate.

Since the goal of most of the early rangeland professionals was to provide high-quality livestock forage, the techniques and systems they developed for rangeland assessments concentrated on the effects of livestock grazing on forage production. The first formal attempt to develop a scientific rangeland survey method was made by James L. Jardine on the Conconino National Forest in 1910 (Chapline and Campbell, 1944).

Early Development of Survey Methods

Jardine's range reconnaissance method involved a careful visual examination of the rangeland to provide a written record of the rangeland's resources. He recorded the following data: (1) a topographic map showing watering places, roads, fences, and cabins; (2) a classification of the rangeland into 1 of 10 grazing or vegetation types; (3) the percentage of the rangeland covered by each forage species; (4) a descriptive report of each grazing or vegetation type, including the suitability of each type for each kind of grazing animal; (5) a map of the timber; and (6) samples of the major species present on the rangeland (Jardine and Anderson, 1919).

Jardine's survey method was highly credible in its time, but it had several shortcomings regarding forage availability estimates. For example, it was based on estimates of the ground cover of each species rather than on direct measurements of the volume or weight of the forage produced by each plant species, and it therefore did not give an accurate measurement of productivity or yield.

Standardization of Rangeland Surveys

In 1933, Standing introduced the concept of using measured volumes of vegetation rather than visual estimates of cover (Standing, 1933). During the 1930s, other modifications were made to the Jardine method, and these were finally standardized as the interagency range survey technique used by the Bureau of Land Management (BLM) and USFS. Although more quantitative than the original reconnaissance method, the interagency survey depended heavily on palatability factors and other subjective criteria for estimating forage production or carrying capacity. This method assessed, almost exclusively, forage production and livestock carrying capacity. Few if any data were collected on soil conditions, wind and water erosion, or other factors that would allow a more comprehensive evaluation of rangelands. More important, the method was not linked to any theoretical base that suggested how the forage composition data that were collected could be interpreted as indicators of ecological conditions on rangelands. Forage production, rather than the state of rangeland ecosystems, was evaluated.

New Theoretical Foundation for Rangeland Surveys

At the same time that Jardine was developing his method for evaluating rangelands, ecologists were developing theories of community dynamics (how plant communities develop and change) that would provide the foundation for new methods for evaluating rangelands.

SUCCESSION AND CLIMAX COMMUNITIES

F. E. Clements of the University of Nebraska, Lincoln, was extremely influential in the study of succession in the Great Plains grasslands. His numerous publications on plant succession and ecology formed a major source of information for resource managers. The textbook *Plant Ecology,* which Clements wrote with his colleague J. E. Weaver (Weaver and Clements, 1938), became a standard in the field. Students from the "Nebraska school of ecology" such as F. W. Albertson, E. J. Dyksterhuis, A. W. Sampson, and L. A. Stoddart became leaders in the young science of rangeland management and brought the Clementsian model of community change into the new field. The Clementsian model dominated much of the early literature in the field.

Clements developed a theory of vegetation dynamics and a quantitative method to test his theory. To Clements, the climax theory rested on the assumption that vegetation could be classified into formations that represented a group of plant species that acted together as if they were a single organism. He wrote, "As an organism, the formation arises, grows, matures, and dies. . . . each climax formation is able to reproduce itself, repeating with essential fidelity the stages of its development" (Clements, 1916:3).

The climax formation was "the climax community of a natural area in which the essential climatic relations are similar or identical" (Clements, 1916:126). (A climax community is the assemblage of plant species that most nearly achieves a long-term steady state of productivity, structure, and composition on a given site [Tueller, 1973].) Clements believed that all successional units within a climatic region developed along one linear path toward a plant community climax that was determined by climate (a climatic climax community). Thus, within a climatic region, a group of plant species would be identified as the climax vegetation, and all sites within that region could be compared with the climax plant species to determine where in the successional path the site was. This theory of vegetation dynamics has been referred to as the *monoclimax theory*.

Clements' method of vegetation analysis involved the use of permanently located quadrats (a plot, usually rectangular, used for ecological and population studies). The species of vegetation in the quadrat was carefully plotted on a map. Changes in vegetation were determined by periodically replotting on a map the species that were present.

The concept of successional change in rangeland ecosystems was to become the fundamental basis of the methods used today to inventory and classify rangelands. Rangelands would be classified on the basis of differences in climax plant community composition and assessed on the basis of the divergence of the current plant composition from the climax plant community composition.

SUCCESSIONAL STAGES AND RANGELAND ASSESSMENT

Sampson (1917) provided what was perhaps the first published reference on the utility of successional stages in rangeland assessment. Then, in 1923, Sampson wrote about the need to move from the old method of determining grazing capacity, which used palatability factors and visual estimates of forage composition, to a new method based on observation of the succession of conspicuous vegetation, that is, the replacement of one set or type of plants by another (Sampson, 1923).

Sampson studied community development in the Watasch Mountains in Utah and classified four developmental stages: the climax herbaceous stage, the mixed grass and weed stage, the late weed stage, and the early weed stage. Although Sampson acknowledged that the Watasch Mountain climax species were not found everywhere, he noted that the character of growth and the habitat requirements of the plants of the different stages were generally the same on native pasturelands. In describing forage production during these four stages, he noted that the climax and the mixed grass and weed stages produced the most forage in terms of quantity and quality (Sampson, 1923).

Sampson noted that the use of successional units to develop a rational grazing plan presumed a detailed knowledge of the successional stages in the development of the vegetation (Sampson, 1923). To obtain this information, he recommended the use of quadrats. However, the great amount of tedious work involved in the mapping and the subsequent synthesis of the data led Sampson to recommend that the person working in the field record the percent cover of all plants of each species within each of the 100 cells that divided the chart quadrat rather than mark the specific location of each plant within each cell. This cover estimate was then multiplied by the palatability of the cover to determine forage yield. Sampson's work was instrumental in bringing successional theory and practical grazing management together.

SUCCESSIONAL STAGES AS CONDITION CLASSES

Sampson's ideas spawned much research into using successional stages as indicators of the status of rangelands. A number of rangeland scientists experimented with methods that could be used to determine the relationship of successional stages to rangeland condition in, for example, Colorado (Hanson et al., 1931), Kansas (Albertson, 1937), Nebraska (Weaver and Fitzpatrick, 1932), North Dakota (Hanson and Whitman, 1938; Sarvis, 1920, 1941), and the intermountain region (Sampson, 1919, 1923).

In 1949, E. J. Dyksterhuis published a landmark paper that was to solidify the contribution of successional theory to the assessment of rangelands. Dyksterhuis refined the climatic climax community described by Clements (1916), proposing that different climaxes coexist as a function of soil or topographic or geographic differences within a similar climate. Dyksterhuis defined those areas that support a unique climax community as a range site. Each site—defined by its climax plant community, soil, and climatic environment—would support a characteristic assemblage of plants, and this vegetation would persist unless it was disturbed by grazing, fire, drought, or other factors. Vegetation would develop toward this climax plant community through successional processes once disturbances (wind, drought, fire) ceased. Grazing drove the plant composition toward the early stages of succession, whereas natural successional processes drove plant composition toward a climax community. By adjusting the grazing pressure or the duration or season of use, rangeland managers could maintain rangelands at any stage of succession. Dyksterhuis proposed a quantitative system for assessing whether a rangeland was at an early or late stage of succession by analyzing the behaviors of three classes of plant species: decreasers, increasers, and invaders. As livestock grazing drove the plant composition toward earlier stages of succession, certain plants were thought to decrease in abundance. These decreasers were replaced by other plants that initially increased in abundance. Those increaser plants were thought to decrease in number and abundance if grazing pushed the plant composition to even earlier stages of succession. The plants that replaced the increasers were called invaders. The successional stage that the rangeland was in could then be determined by what proportion of the vegetation, measured by percent composition by weight, was decreasers, increasers, or invaders. If most of the plants were decreasers, the rangeland was thought to be in a late successional stage; if most plants were invaders, the rangeland was considered to be in a very early stage of succession.

Dyksterhuis also proposed that the condition of rangelands improved as succession progressed. Later successional stages were thought to provide better forage and to be more stable and productive plant communities. The condition of a rangeland could therefore be determined by the climax plant community of the site. The greater the proportion of increasers or invaders, the poorer the condition. The greater the proportion of decreasers, the better the condition.

ADOPTION OF THE SUCCESSION-RETROGRESSION MODEL BY FEDERAL AGENCIES

Dyksterhuis's use of successional stages as the measure of the condition of rangelands had great appeal. His concept not only proposed a systematic way of evaluating the condition of rangelands but also explained the effects of grazing on rangeland vegetation and provided the basis for changes in grazing management. Estimations of livestock carrying capacity were linked to range sites, condition classes, and successional stages. By 1950, the measurement of range condition (Soil Conservation Service [SCS]) as the degree of departure from climax plant community (SCS) vegetation of a defined range site and the succession-retrogression model of rangeland development became the standard concept in U.S. rangeland management. All major inventory and classification methods in use today are modifications of that basic concept.

The concept was adopted to varying degrees by the USFS, BLM, and

SCS, the agencies with the most responsibility for rangeland management in the United States. Changes in terminology and interpretation since 1950 have resulted in divergences between the site classification definitions and the rangeland inventory methods used by the different agencies.

APPLICABILITY OF THE SUCCESSION-RETROGRESSION MODEL

Even as the succession-retrogression model was accepted by rangeland scientists and institutionalized in the federal management agencies (SCS, BLM, USFS), other community ecologists began to question the validity of the concept of climax community itself.

This debate was missing from the first editions of Stoddart and Smith's (1943) and Sampson's (1952) range management textbooks (Smith, 1989) and is still missing from the later textbooks of Stoddart and colleagues (1975) and Heady (1975) and the most recent range research methods book edited by Cook and Stubbendieck (1986). The books and reference papers leading to the development of range site (SCS) and range condition (SCS), however, make no mention of papers by Cain, Egler, or Gleason that questioned the successional model (Smith, 1989).

Within the range science literature, investigators criticized the subjective nature of habitat type (West, 1982) and range site (Laycock, 1989) classifications, yet the Range Inventory Standardization Committee of the Society for Range Management recommended governmentwide use of the ecological site classification system (Society for Range Management, Range Inventory Standardization Committee, 1983), which was based on the same community ecology theories of previous classifications. New developments in community ecology including the analysis of community structure and causal factors done by multivariate techniques (statistical analysis of the interaction of multiple causes of rangeland change) and new community ecology ideas, like the threshold concepts of Friedel (1991), have not been incorporated into the methods used to inventory, classify, or monitor rangelands.

LINKS BETWEEN OTHER BRANCHES OF ECOLOGY
AND RANGELAND SCIENCE

New developments in ecological research have had an influence on specialization and research within the field of range science, but this influence on the diversity of range science research has not yet been transferred to a diversification of the measures used to inventory and monitor rangeland. The fundamental concepts underlying the rangeland classification and inventory methods of all of the federal agencies are based on those proposed by Sampson, Clements, and Dyksterhuis in the first half of the twentieth century.

CURRENT AGENCY RANGELAND ASSESSMENT THEORY AND PRACTICE

All federal agencies measure range condition (SCS) or ecological status (USFS and BLM) as the degree to which the vegetation of a site is different from the climax plant community or potential natural community characteristic of that or similar sites. SCS, USFS, and BLM have adopted systems that use (1) ecological site (BLM) or range site (SCS) as the landscape subdivision on which the analysis is made, (2) climax plant community (SCS) or potential natural community (USFS and BLM) as the standard against which range condition (SCS) or ecological status (USFS and BLM) is judged, and (3) succession and retrogression models as the primary means of explaining the ways that rangelands change.

When and how these agencies use range site (SCS) or ecological site (USFS and BLM) and how a rangeland is given a range condition (SCS) or ecological status (USFS and BLM) rating vary. SCS assists nonfederal landowners with developing and implementing conservation plans to protect soil, water, and other natural resources on their rangelands and conducts the National Resources Inventory. The methods used to classify sites and the soils on those sites are standardized at the national level. The SCS *National Range Handbook* (U.S. Department of Agriculture, Soil Conservation Service, 1976) prescribes the procedures that should be used by agency employees. Each state SCS office standardizes the methods used within that state to evaluate rangelands.

USFS and BLM have responsibility for managing federal lands for multiple uses on a sustained-yield basis and for maintaining a data base of the ecological status (USFS and BLM) of the lands under their jurisdiction. The site classification and evaluation methods used by BLM are similar to those used by SCS and are standardized at the national level. The USFS site classification and evaluation methods have recently been standardized at the national level in newly released manuals (U.S. Department of Agriculture, Forest Service, 1991b).

The responsibility for developing and implementing the ecological type (USFS) classification system in USFS has been assigned to regional foresters and forest and range experiment station directors. These individuals also have responsibility for correlating ecological type (USFS) descriptions across regional boundaries and between and among other agencies (U.S. Department of Agriculture, U.S. Forest Service, 1991b).

Rangeland Reference Terms

Three federal agencies (the Bureau of Land Management [BLM], the U.S. Forest Service [USFS], and the Soil Conservation Service [SCS]) evaluate and classify rangelands, but the techniques and evaluation criteria vary somewhat. The following is a comparative analysis of the definitions each agency uses.

Terms of Reference	Agency	Definition
Status Rating		
Ecological status	BLM, USFS	Four classes used to express the degree to which the makeup of the present vegetation reflects the potential natural community. The class or rating, percentage of vegetation present in a potential natural community, is as follows: potential natural community, 76 to 100 percent; late seral, 51 to 75 percent; midseral, 26 to 50 percent; early seral, 0 to 25 percent.
Range condition	SCS	Four classes used to express the degree to which the makeup of the present vegetation reflects the climax plant community. The class or rating, percentage of vegetation present in a climax plant community, is as follows: excellent, 76 to 100 percent; good, 51 to 75 percent; fair, 26 to 50 percent; poor, 0 to 25 percent.
Site Classification		
Range site	SCS	A distinctive kind of rangeland that differs from other kinds of rangelands in its ability to produce a characteristic natural climax plant community.
Ecological type	USFS	A category of land with a unique combination of potential natural community, soil, landscape features, and climate; it differs from other ecological types in its ability to produce vegetation and respond to management.

Terms of Reference	Agency	Definition
Site Classification—continued		
Ecological site	BLM	A kind of land with a specific potential natural community and specific physical site characteristics; it differs from other kinds of land in its ability to produce vegetation and respond to management.
Trend	SCS, USFS	*Trend* is described as up, down, or not apparent (also static or stable). *Up* represents a change toward a climax of potential natural community; *down* means a change away from a climax or potential natural community; *not apparent* means there is no recognizable change.
Apparent trend	SCS, BLM	A judgment of trend based on a one-time observation. It includes consideration of such factors as plant vigor; abundance of seedlings and young plants; accumulation or lack of plant residues on the soil surface; and soil surface characteristics including crusting, gravel pavement, pedestaled plants, and sheet or rill erosion.
Benchmark Plant Communities		
Climax plant community	SCS	The natural plant community that would be found on a range site in the absence of abnormal disturbances and physical site deterioration. It includes only native plant species.
Potential natural community	USFS, BLM	The biotic community that would become established if all successional sequences were completed without interferences by humans under the present environmental conditions. It may include naturalized nonnative species.

Site Classification

SCS uses the term "range site" to classify different rangelands. BLM has adopted site classifications that are similar in concept to those of SCS. USFS uses the term "ecological type" to classify its rangelands. All three agencies classify rangelands into different types on the basis of the kinds and amounts of plants expected in the climax plant community (SCS) or potential natural community (USFS and BLM) thought to be characteristic of that type of rangeland.

RANGE SITE CLASSIFICATION

SCS pioneered the use of range sites for rangeland classification and has used the concept fairly consistently since the 1940s. The SCS *National Range Handbook* (U.S. Department of Agriculture, Soil Conservation Service, 1976) defines a range site (SCS) as a specific area that is "capable of supporting a native plant community typified by an association of species that differs from that of other rangeland sites in the kind or proportion of species or in total production" (Section 302.1). Since a range site (SCS) is defined as a native plant community, no introduced or exotic species can be considered part of the climax plant community used to define the range site (SCS). This restriction of the definition to native plants is the only major difference that distinguishes range site from the site classifications used by USFS and BLM.

SCS considers each site to be the product of all the environmental factors responsible for its development, including soils, vegetation, topography, climate, and fire. Soil surveys are particularly useful in site classification. Soil surveys classify soils with similar properties into mapping units. The characteristics that are used to classify soils are primarily those that can be measured in the field such as color, arrangement of horizons, pH, texture, and other morphological features. From knowledge of the mapping unit that contains the soil being classified, SCS soil scientists can estimate such soil attributes as the depth of the soil influenced by organic matter, the mineral or chemical content of soil horizons, the rate at which the soil takes up water, the water storage capacity of the soil, the soil's vulnerability to erosion, and the soil's fertility. The differences in soil properties described in soil surveys are important elements in classifying rangelands into range sites (SCS).

For each site, the climax plant community composition is defined as that which existed before human influence. In the absence of abnormal disturbances that upset ecological processes or physical deterioration of the site, the unique interaction on a given site is thought to support a plant community characterized by plant species that differ from those in the plant community found on another site in terms of the kind or proportion of species or total annual vegetative production.

Range sites are mapped and correlated by matching soils and climate zones with what is determined to be the characteristic climax vegetation. Data for this purpose are derived from many sources, including the following:

- evaluation of the vegetation and soils on rangelands that have been protected from disturbance for long periods of time;
- comparison of areas that are used to various degrees by livestock with similar ungrazed areas;
- evaluation and interpretation of research dealing with natural plant communities and soils;
- review of early historical and botanical literature; and
- prediction of climax vegetation on the basis of information gathered from areas with similar soils and climates.

SOILS AND CLIMATE   Soils and climates that result in the same climax vegetation and annual biomass production are considered to make up the same range site (SCS). Plant composition, measured by the weight of biomass produced annually by each species, is the key descriptor of a range site (SCS). The range site (SCS) usually remains the same as long as the soil and climate remain unchanged. If the soil has been changed by erosion or some other factor so that the changed soil in combination with the site's climate cannot support the growth of the characteristic climax vegetation, then a new range site (SCS) must be defined.

Individual sites are identified and differentiated from others on the basis of specific criteria. The criteria are

(1) significant differences in the species or species groups that are ecological dominants in the plant community; (2) significant differences in the proportion of species or species groups that are ecological dominants of the plant community; and (3) significant differences in the total annual production of the plant community (U.S. Department of Agriculture, Soil Conservation Service, 1976:Section 302.6).

Table 3-1 provides a comparison of the vegetation found on three different soil types in the same county in Utah and illustrates how plant community data are used to arrive at rangeland site definitions (Shiflet, 1973). All three soils support essentially the same plant community in terms of species composition. All sites are dominated by bluebunch wheatgrass (*Agropyron spicatum*), with only minor differences in the other components. However, there was a statistically significant difference in total production between the Manila and the Broad soils, with the Manila soil being approximately 17 percent more productive than the Broad soil.

TABLE 3-1 Average Production and Composition of Vegetation Produced on Three Soil Types in Box Elder County, Utah

Major Species	Middle Soil[a] Production (kg/ha)	Middle Soil[a] Composition (percent)	Broad Soil[b] Production (kg/ha)	Broad Soil[b] Composition (percent)	Manila Soil[c] Production (kg/ha)	Manila Soil[c] Composition (percent)
Bluebunch wheatgrass (Agropyron spicatum)	1,650*	85	1,462†	82	1,833*	88
Sandberg bluegrass (Poa secunda)	32	2	58	3	24	1
Balsamroot (Balsamorhiza sagittata)	50	3	34	2	7	T
Cheatgrass (Bromus tectorum)	9	T	7	T	1	T
Yellowbrush (Chrysothamnus viscidiflorus var. lanceolatus)	32	2	64	4	36	2
Big sagebrush (Artemisia tridentata)	29	1	32	2	30	1
Bitterbrush (Purshia tridentata)	44	2	—	—	1	T
Other species	92	5	116	7	140	8
Total	1,938*,†	100	1,773†	100	2,072*	100

NOTE: Data are based on 30 observations of 10 subplots each. Reading across columns, production values followed by the same symbol (* or †) were not significantly different at the 5 percent probability level. T, trace percentage (less than 0.5 percent); —, did not occur or does not apply.

[a]Loamy-skeletal, mixed, mesic Calcic Haploxeroll.
[b]Loamy-skeletal, mixed, frigid Calcic Argixeroll.
[c]Fine, Montmorillonitic, frigid Typic Argixeroll.

SOURCE: Adapted from T. N. Shiflet. 1973. Range sites and soils in the United States. Pp. 26–33 in Arid Shrublands: Proceedings of the Third Annual Workshop of the United States/Australia Rangeland Panel, D. H. Hyder, ed. Denver: Society for Range Management.

Assuming that a 15 percent difference were large enough to affect grazing management, the plant community that grows on Manila soils would be classified as a separate site.

An example of how differences in soils, production, and plant composition interact to influence the determination of rangeland site can be seen in three rangelands that were studied for 10 years in a project described by Williams and Hugie (1966) (Table 3-2). The only major difference in the environments of the three study locations was the soil. Production from the Hoelzle and Bancroft soils did not differ significantly. On average, however, the plant communities on these soils were approximately 35 percent more productive than those that grew on Goodington soils. This difference was statistically significant and large enough to require different management practices (Shiflet, 1973). The plant community that grew on Goodington soil would therefore be separated from those that grew on the other two soil types because of its lower productivity. Plants that grew on the Hoelzle and Bancroft soils did not differ in production, but they did differ in species composition. Idaho fescue (*Festuca idahoensis*) was a significantly higher producer in the plant community that grew on Hoelzle soils than it was in the community that grew on Bancroft soils, even though it was the most important herbaceous species on both soils. Another difference in the two plant communities was within the shrub component. Big sagebrush (*Artemisia tridentata*) was the major shrub in the community growing on Hoelzle soil but did not occur at all in the plant community growing on Bancroft soils. On the other hand, three-tip sagebrush (*Artemisia tripartita*) accounted for only 2 percent of the production of the vegetation on Hoelzle soils but was the most important shrub on the Bancroft soils, contributing 17 percent of the total production. On the basis of the lower productivity of the plant community growing on Goodington soils and differences in species composition and proportion of species between the communities growing on the Hoelzle and Bancroft soils, Williams and Hugie concluded that all three were unique and represented three distinct range sites (SCS) (Shiflet, 1973).

TOPOGRAPHY   Topography, too, can play an important role in distinguishing between range sites (SCS). Features such as the slope of the land, the direction the sloping land faces and whether the land is located at the top or bottom of the slope affect runoff and delivery of water, evaporation, temperature, and other factors that influence the kinds and amounts of plants that grow on a site. Often, soils that are quite similar in many respects support different vegetation because of the influence of topography and are classified as different range sites (SCS). Table 3-3 illustrates the effects of exposure on two study areas located on opposite sides of a hill in southern Idaho. The soils on the two sites were very similar. There were only minor differences in the structures of the two plant communi-

TABLE 3-2 Average Production and Composition of Vegetation Produced on Three Soils in Blaine County, Idaho

Major Species	Goodington Soil[a]		Hoelzle Soil[b]		Bancroft Soil[c]	
	Production (kg/ha)	Composition (percent)	Production (kg/ha)	Composition (percent)	Production (kg/ha)	Composition (percent)
Idaho fescue (Festuca idahoensis)	187	24	430	39	257	24
Bluebunch wheatgrass (Agropyron spicatum)	62	8	121	11	159	15
Sandberg bluegrass (Poa secunda)	144	18	149	13	91	9
Bottlebrush squirreltail (Sitanion hystrix)	94	12	41	4	—	—
Prairie June grass (Koelaria cristata)	13	2	6	1	84	8
Narrowleaf pusseytoes (Antennaria stenophylla)	43	5	—	—	—	—
Longleaf phlox (Phlox longifolia)	31	4	19	2	26	
Hawksbeard (Crepis acuminata)	—		9	1	47	4

Balsamroot (Balsamorhiza sagittata)	—	—	8	1	—	—
Other annuals	55	7	38	3	11	1
Big sagebrush (Artemisia tridentata)	25	3	67	6	—	—
Three-tip sagebrush (Artemisia tripartata)	11	1	21	2	175	17
Desert rabbitbrush (Chrysothamnus viscidiflorus)	14	2	17	2	27	3
Other species	114	14	177	15	166	16
Total	793*	100	1,103†	100	1,043†	100

NOTE: Data are based on 10 annual observations of 20 subplots each. Reading across columns, production values followed by the same symbol (* or †) were not significantly different at the 5 percent probability level. T, trace percentage (less than 0.5 percent); —, did not occur or does not apply.

[a]Fine, Montmorillonitic, frigid Typic Palexeroll.
[b]Has been eliminated as a separate soil series.
[c]Fine-silty, mixed, frigid Calcic Argixeroll.

SOURCE: Adapted from T. N. Shiflet. 1973. Range sites and soils in the United States. Pp. 26–33 in Arid Shrublands: Proceedings of the Third Annual Workshop of the United States/Australia Rangeland Panel, D. H. Hyder, ed. Denver: Society for Range Management.

TABLE 3-3 Forage Production, Composition, and Frequency of Vegetation Produced on Two Exposures of a Silt Loam (Trevino soil[a]) in Power County, Idaho

Major Species	South Exposure			North Exposure		
	Production (kg/ha)	Composition (percent)	Frequency (percent)	Production (kg/ha)	Composition (percent)	Frequency (percent)
Bluebunch wheatgrass (*Agropyron spicatum*)	165*	27	84	172*	24	100
Sandberg bluegrass (*Poa secunda*)	99*	17	100	119*	17	100
Thurber needlegrass (*Stipa thurberiana*)	77*	12	100	62*	9	100
Balsamroot (*Balsamorhiza sagittata*)	43*	7	12	78*	11	29
Hawksbeard (*Crepis acuminata*)	39*	6	38	53*	7	51
Longleaf phlox (*Phlox longifolia*)	20	3	83	26	4	88
MacDougal lomatium (*Lomatium macdougali*)	17	3	62	1	1	1

Nineleaf lomatium (Lomatium triternatum)	1	T	62	11	8	132
Cheatgrass (Bromus tectorum)	19	3	64	3	T	23
Other annuals	1	T	1	2	T	2
Big sagebrush (Artemisia tridentata)	95*	15	28	124*	17	36
Bitterbrush (Purshia tridentata)	1	T	1	6	1	20
Desert rabbitbrush (Chrysothamnus viscidiflorus)	—	—	—	4	4	5
Other species	41	7	63	8	—	—
Total	618*	100	—	721†	100	—

NOTE: Data are based on 10 annual observations of 20 subplots each. Reading across columns, production values followed by the same symbol (* or †) were not significantly different at the 5 percent probability level. T, trace percentage (less than 0.5 percent); —, did not occur or does not apply.

[a]Loamy, mixed, mesic, Lithic Xerollic Camborthid.

SOURCE: Adapted from T. N. Shiflet. 1973. Range sites and soils in the United States. Pp. 26–33 in Arid Shrublands: Proceedings of the Third Annual Workshop of the United States/Australia Rangeland Panel, D. H. Hyder, ed. Denver: Society for Range Management.

ties. There was a significant difference in total production, however, with the site exposed to the north producing an average of 17 percent more biomass than that produced on the site exposed to the south. The two exposures were classified as different sites (Shiflet, 1973).

ECOLOGICAL SITES AND TYPES

The site classifications proposed for use by BLM for rangeland classification are similar to those used by SCS. They differ primarily in terminology rather than concepts (Society for Range Management, Range Inventory Standardization Committee, 1983). BLM's system is based on ecological sites (BLM) that would be expected to produce a characteristic potential natural community (USFS and BLM) that has a predictable plant composition and annual production. Potential natural community (BLM) describes a plant community composition that accepts some naturalized nonnative species in that community. This differs from the SCS definition of climax plant community (SCS) (see above).

The USFS is changing its classification of rangelands. It uses the term "ecological type" (USFS) to classify its rangelands. The term is defined in a newly released USFS manual as "a category of land having a unique combination of potential natural community, soil, landscape features, climate, and differing from other ecological types in its ability to produce vegetation and respond to management" (U.S. Department of Agriculture, U.S. Forest Service, 1991b:Section 2090.11-05). Descriptions of vegetation (potential natural community [USFS and BLM] and successional stages), soils, topographic features, water, climate, geology, and management interpretations are to be included in ecological type (USFS) descriptions.

The proposed USFS classification system is similar to the SCS system, but it differs in the definition of potential natural community (USFS and BLM). USFS defines potential natural community as follows: "The biotic community that would be established if all successional sequences of its ecosystem were completed without additional human-caused disturbances under present environmental conditions. Grazing by native fauna, and natural disturbances, such as drought, floods, wildfire, insects, and disease, are inherent in the development of potential natural communities which may include naturalized non-native species" (U.S. Department of Agriculture, U.S. Forest Service, 1991b:Section 2090.11-05).

Descriptors of potential natural community (USFS and BLM) include, at a minimum, (1) a list of plant species on the site; (2) some measures of plant species composition or the dominant plant species; (3) production parameters such as weight, cover, basal area (the cross-sectional area of plant stems), incremental growth, or site index (an indicator of site productivity based on measuring the height to which a tree on the site has grown in 50 years); (4) a measure of constancy (an indication of the likelihood of finding a species in a given community), by species; and (5) general environmental data. "Ideally, a well described [potential natural community] will show all communities within the sere [a sere is one of a series of ecological communities formed in ecological succession] to be expected following different kinds of disturbance" (U.S. Department of Agriculture, U.S. Forest Service, 1991b:Section 2090.11-2.15).

Evaluation of Range Condition and Ecological Status

SCS, USFS, and BLM evaluate successional change on rangelands by comparing the composition and annual biomass produced by the existing vegetation with a previously determined benchmark plant composition and production. This benchmark is defined by SCS as the climax plant community (SCS) for that range site (SCS), and it is defined by USFS and BLM as the potential natural community (USFS and BLM) for that ecological type (USFS) or ecological site (BLM), respectively. SCS defines the term "range condition" (SCS) as "the present state of vegetation of a range site in relation to the climax plant community for that site" (U.S. Department of Agriculture, Soil Conservation Service, 1976:Section 305.2). USFS defines the term "ecological status" (USFS) as the "degree of similarity between existing vegetation (all components and characteristics) and soil conditions compared to the Potential Natural Community and the desired soil condition on a site" (U.S. Department of Agriculture, U.S. Forest Service, 1991b:Section 2090.11-05). BLM also uses the term "ecological status" (USFS and BLM) to describe "the present state of vegetation in relation to the potential plant community" (U.S. Department of the Interior, Bureau of Land Management, 1985a:Section). All current systems depend on a comparison of the current plant composition with an established benchmark plant composition.

RANGE CONDITION (SCS)

The range condition (SCS) on a given site is based on concepts of succession and retrogression toward or away from the defined climax plant community (SCS). That is, the climax species composition of a particular range site (SCS) changes because of grazing, climatic cycles, fire, insects, physical disturbances, and other stresses (U.S. Department of Agriculture, Soil Conservation Service, 1976). These stresses result in the replacement of species that are characteristic of the climax plant community (SCS) composition by other species characteristic of earlier successional stages. Rangeland managers measure the species composition of a site and compare it with that expected in the climax plant community (SCS). The similarity is the range condition (SCS). A retrogression (decline) in range condition (SCS) is assumed to be a predictable change from one set of species to another when the same disturbance (for example, grazing or drought) is working. Likewise, the recovery of a site is assumed to be a predictable process of succession. Plants that grow on the deteriorated area are replaced by species that are typical of a later stage of succession until the climax plant community (SCS) is ultimately restored.

Range condition (SCS) classification was originally designed to measure the influence of livestock grazing on the composition and production of the plant community. The hypothesis was as follows:

> Grazing management determines place, time, and amount of foliage removal. Removal of green foliage by grazing retards growth most among the species grazed most. This favors the species grazed least because more of the water, nutrients, and light available per unit of surface is left for them. Thus, without practical management of grazing, the normally shorter species are favored as well as the least palatable and the annual species. As the taller species lose ground under close grazing, their place is taken by species short enough to escape with a high percentage of their foliage ungrazed. The result is general reduction in yield as well as measurable change in species composition.
>
> On rangelands, this process, fortunately, can be reversed. Ordinarily a change in management of grazing is all that is required. If secondary plant succession is permitted or fostered, the combination of plants that produces the greatest tonnage of foliage will crowd out other combinations of plants resulting from past mismanagement. Foliage on rangeland in top condition is almost all forage (Dyksterhuis, 1958:151).

The link between grazing pressure and condition class was also thought to be clear and direct.

> It follows that the range condition classification provides a measure of determining and stating specifically *how much* a pasture is overgrazed. In the range-condition system the amount of improvement possible is the exact reciprocal of the amount overgrazed. . . .
>
> Accordingly, it is more informative to name the condition class than to point out overgrazed areas. Excellent range condition means the site is not overgrazed, while good, fair, and poor conditions refer to three specific grades of past overuse that show today, but can be corrected through one, two, or three classes (Dyksterhuis, 1958:154).

To determine the range condition (SCS) of a site, SCS measures the amount (in kilograms per hectare or pounds per acre) of annual production (air dry weight) by clipping and weighing the biomass or by using standardized procedures to estimate the amount of the current year's plant growth in randomly located plots (U.S. Department of Agriculture,

Soil Conservation Service, 1976). The clipped or estimated weights determined for individual plant species are divided by the total weight of all species to determine the relative percent composition of each species. The percentage of each species is compared with the percentage expected in the climax plant community (SCS) to determine range condition (SCS). Standardized forms are provided by each state SCS office for use in making these calculations and recording the ratings. The forms provide employees with the option of recording other information about the particular site, for example, erosion treatment needs and special considerations such as riparian areas.

Sites occupied by 76 to 100 percent of their climax plant community (SCS) species are rated in excellent condition. Those occupied by 51 to 75 percent of the climax plant community (SCS) species are rated as good, those with by 26 to 50 percent of the climax plant community are rated as fair, and those with 25 percent or less of their climax vegetation are rated as poor (U.S. Department of Agriculture, Soil Conservation Service, 1976).

Although the species composition information described above is the most important factor in determining range condition (SCS), SCS rangeland conservationists can lower the condition rating if certain conditions are present. These conditions include the following:

- if the overall productivity of the plants on the site is lower than normal and cannot be explained by abnormal conditions (for example, drought);
- if many of the plant species expected to be found on the site are missing; and
- if there are signs of accelerated erosion (U.S. Department of Agriculture, Soil Conservation Service, 1976).

Table 3-4 shows how range condition (SCS) can be determined for a particular range site (SCS). The table lists the major species expected in the climax plant community (SCS) characteristic of the Middle Cobbly Loam range site (SCS) in Box Elder County, Utah. Cheatgrass (*Bromus tectorum*) and snakeweed (*Gutierreza sarothrae*) are not considered part of the climax plant community (SCS) for the Middle Cobbly Loam range site (SCS), but they were found on the site being rated, so they are listed as having zero potential composition. The potential species composition is the benchmark against which the site to be rated is compared. The estimated percent composition of each species found during the investigation of the site can then be used to calculate the range condition (SCS) rating. This comparison of the actual plant composition with the composition of the climax plant community (SCS) is illustrated in Figure 3-1.

Allowable composition can be equal to, but cannot exceed, the potential percent composition for each species expected in the climax plant

TABLE 3-4 Determination of a Range Condition (SCS) Rating for Middle Cobbly Loam Range Site (SCS) in Box Elder County, Utah

Major Species	Climax Plant Community		Rated Rangeland Community		
	Potential Production (kg/ha)	Potential Composition (percent)	Estimated Production (kg/ha)	Composition (percent)	Allowable Composition (percent)
Bluebunch wheatgrass (Agropyron spicatum)	1,650	85	701	45	45
Sandberg bluegrass (Poa secunda)	32	2	137	9	2
Balsamroot (Balsamorhiza sagittata)	50	3	18	1	1
Big sagebrush (Artemisia tridentata)	29	1	262	17	1
Bitterbrush (Purshia tridentata)	44	2	208	14	2
Yellowbrush (Chrysothamnus viscidiflorus var. lanceolatus)	32	2	6	T	T
Cheatgrass (Bromus tectorum)	0	0	136	8	0
Snakeweed (Gutierreza sarothrae)	0	0	34	2	0
Other species	92	5	58	4	4
Total	1,929	100	1,526	100	55

NOTE: T, trace percentage (less than 0.5 percent).

SOURCE: Adapted from T. N. Shiflet. 1973. Range sites and soils in the United States. Pp. 26–33 in Arid Shrublands: Proceedings of the Third Annual Workshop of the United States/Australia Rangeland Panel, D. H. Hyder, ed. Denver: Society for Range Management.

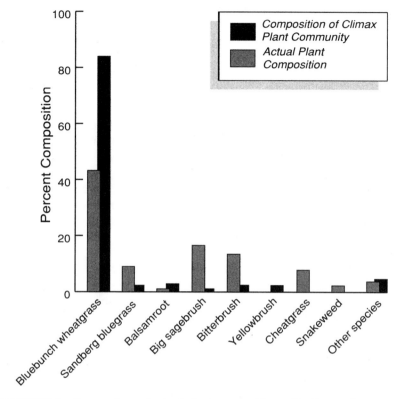

FIGURE 3-1 Comparison of actual plant composition with climax plant community composition for range condition (SCS) rating. Source: Derived from T. N. Shiflet. 1973. Range sites and soils in the United States. Pp. 26-33 in Arid Shrublands: Proceedings of the Third Annual Workshop of the United States/Australia Rangeland Panel, D. H. Hyder, ed. Denver: Society for Range Management.

community (SCS). The potential composition of bluebunch wheatgrass (*Agropyron spicatum*), for example, is 85 percent, and the estimated composition of bluebunch wheatgrass on the site being rated is 45 percent; therefore, 45 percent is recorded as an allowable composition for bluebunch wheatgrass in Table 3-4. On the other hand, the potential composition of sandberg bluegrass (*Poa secunda*) is 2 percent, but it was estimated at 9 percent on the rangeland being rated. Since the allowable composition cannot exceed the potential composition, only 2 percent is recorded for the allowable composition of sandberg bluegrass in Table 3-4. Cheatgrass and snakeweed were estimated to be present at 8 and 2 percent compositions, respectively. Neither species, however, is considered part of the climax plant community (SCS) for the Middle Cobbly Loam range site so both species are given 0 percent allowable percent composition.

This process is followed until all of the species found on the site being rated are assigned an allowable percent composition. The total of the allowable percent composition of each species is then used to determine the range condition (SCS) rating. In the example presented in Table 3-4, the range condition (SCS) rating is 55, meaning that the current composition of the rangeland is 55 percent of its potential climax plant community (SCS) composition. The rangeland is then assigned to a range condition class on the basis of its range condition (SCS) rating as described above. The range condition of the rangeland described in Table 3-4 would be good. The range condition (SCS) class of the Middle Cobbly Loam range site could be lowered from good to fair if the investigator noted and recorded signs of serious erosion or if total production was substantially lower than expected for the Middle Cobbly Loam range site in good range condition (SCS).

The SCS office in each state has developed a form to be used by its personnel when doing a range condition (SCS) rating. (See the Appendix for an example of such a form.) The appearance of the form may differ from state to state, but the procedure is standardized.

ECOLOGICAL STATUS (USFS AND BLM)

The BLM system measures the departure from the potential natural community (USFS and BLM) plant composition and production in much the same way that SCS measures the departure from the climax plant community (SCS) composition (U.S. Department of the Interior, Bureau of Land Management, 1985a). An ecological status (USFS and BLM) rating, which is calculated as a percentage of the potential natural community composition and production, is calculated on the basis of field measurements of plant composition and production by weight. Ecological status (USFS and BLM) ratings are reported in the following four classes:

1. Early seral—0 to 25 percent of the potential natural community (USFS and BLM) is being produced.

2. Midseral—26 to 50 percent of the potential natural community (USFS and BLM) is being produced.

3. Late seral—51 to 75 percent of the potential natural community (USFS and BLM) is being produced.

4. Potential natural community (USFS and BLM)—76 to 100 percent of the potential natural community (USFS and BLM) is being produced (U.S. Department of the Interior, Bureau of Land Management, 1985a).

These ecological status (USFS and BLM) ratings are based on the departure of the present plant community from the potential natural community. As such, the status is a reflection of current vegetation in relation to the established potential natural community (USFS and BLM) for that ecological site (BLM).

USFS uses ecological status (USFS and BLM) as its measure of the degree of similarity between the existing vegetation and soil conditions compared with those of the potential natural community and the desired soil condition on a site. This similarity can be expressed on a relative scale ranging from 0 to 100, with adjectival ratings assigned as low, moderate, or high similarity. USFS methods of determining ecological status (USFS and BLM) are similar in approach to those of SCS. The frequency of occurrence, however, rather than an estimate of production from clipped weights is sometimes used as the measure of plant composition (U.S. Department of Agriculture, U.S. Forest Service, 1991b).

Trend and Apparent Trend

The concept of trend (a change in a certain characteristic of a rangeland over time) has been used since the early 1900s. It provided a simple, easily applied, and cost-effective method for determining whether grazing practices were causing the desired change in rangelands. Increased soil cover might, for instance, be a goal of management. After the correct stocking rate is calculated, a baseline measure of vegetative cover is made. Cover measurements are made on the same area on a yearly or other established schedule. A downward trend usually results in further adjustments in the level or timing grazing allowed on the site.

Interpretation of the trend concept as a measure of change in range condition (SCS) or ecological status (USFS and BLM) became standardized and accepted in rangeland inventorying and monitoring. A trend in either range condition (SCS) or ecological status (USFS and BLM) is now strictly defined as an evaluation of whether the plant composition and production of a particular rangeland is becoming more similar or less similar to the defined climax plant community (SCS) or potential natural community (SCS) for that particular rangeland site (or ecological type [USFS] or ecological site [BLM]).

Trend is determined by repeated evaluation of range condition (SCS) or ecological status (USFS and BLM) at the same location but at different points in time. Managers, however, often needed a more immediate assessment of trend to evaluate the effect of a change in management. Range scientists attempted to develop a set of criteria that would indicate a trend in ecological state from an evaluation at one point in time. This evaluation of trend has been referred to as *apparent trend* (SCS).

SCS lists several characteristics of vegetation and soil that can be used to indicate apparent trend, including the abundance of later-succession (desirable) seedlings and earlier-succession (undesirable) young plants, accumulation of plant residues, plant vigor, and the condition of the soil surface.

According to SCS, the relative importance of these trend indicators varies depending on the site's vegetation, soils, and climate. Any single indicator will give some indication of change in range condition (SCS), but SCS recommends that the evaluation of apparent trend (SCS) be made by considering all indicators in their proper relation to each other (U.S. Department of Agriculture, Soil Conservation Service, 1976).

Apparent trend (SCS) is a professional judgment made at one point in time. It is based on the current status of the vigor, reproduction, distribution, age, and other characteristics of the vegetation as well as on soil, litter cover, erosion, and other factors of a site.

NEW METHODS NEEDED TO ASSESS
RANGELAND HEALTH

The current range condition (SCS) or ecological status (USFS and BLM) ratings systems have four components: (1) the classification of rangelands into sites based primarily on differences in their expected climax plant communities (SCS) or potential natural communities (USFS and BLM), (2) the rating of condition or status as the degree to which the current plant composition and production represent those of the benchmark climax plant community (SCS) or potential natural community (USFS and BLM), (3) estimation of site potential on the basis of the plant composition and production estimated for the defined climax plant community (SCS) or potential natural community (USFS and BLM), and (4) estimation of whether a rangeland is changing in ways that will make it more or less similar to the defined climax plant community or potential natural community (USFS and BLM).

The current system of rangeland assessment will not serve as an adequate evaluation of rangeland health, as defined by the committee. The current system does not adequately assess soil stability or the integrity of ecological processes such as nutrient cycles and energy flow. There are problems with each of the four components of the current system that will limit its utility as a measure of rangeland health.

Site Classification

Even the casual observer can look at a rangeland and recognize that some parts of the landscape are different from others. These differences are expressed in many ways, but they are expressed most obviously through topography, vegetation, and soil surface characteristics like rocks, bare soil, or litter cover. In practice, the resulting site delineations are arbitrary landscape divisions that can be used to provide order to a complex system for management purposes and extrapolation of research results (Passey and Hugie, 1962; Pendelton, 1989; Tueller, 1973). The major use of the site concept in rangeland assessment has been for analysis of range condition (SCS) or ecological status (USFS and BLM). Sites therefore have most often been delineated by studying the composition and production of climax community plant species (Shiflet, 1973).

CLASSIFICATION BASED ON CLIMAX CONCEPT

A focus on climax plant community (SCS) or potential natural community (USFS and BLM) composition for distinguishing between sites is the common point of departure for all site classification systems used by federal agencies. Differences in the expected plant composition and production in the climax plant community have been used as a way of distinguishing the variability in plant composition caused by use, management, stage of succession, or other variables from the variability caused by differences in the site.

The reliance on differences in the expected climax plant community (SCS) or potential natural community (USFS and BLM) composition for distinguishing between sites, however, has been questioned on both theoretical and practical grounds. Plant composition at any one point in time varies because plant communities are constantly changing in composition and production owing to changes in environmental influences (Gleason, 1926; Harper, 1977). Some scientists have questioned whether the concept of a single, definable, and predictable climax plant community can be applied to all rangelands (West, 1985; Wilson, 1989). Others have suggested that succession may follow multiple pathways and that the pathway followed by a particular rangeland depends on the kind of disturbance and the environmental conditions during secondary succession (Friedel, 1991; Risser, 1989; Smith, 1989; Westoby et al., 1989). The plant community that arises from primary succession (succession that begins on a habitat that has not previously been inhabited), therefore, may be different from the plant community that arises from secondary succession (succession that follows destruction of all, or part, of a previous plant community). Differences in expected successional processes and their resulting end points may be due as much to the type of disturbance and the prevailing conditions at the time of disturbance as to differences between sites.

The practical difficulties of determining climax vegetation have also been cited. Although there may not be relic (undisturbed) rangelands that can be used as benchmarks of climax plant community composition (Shiflet, 1973), several methods can be used to reconstruct a hypothetical climax plant community: (1) interpolation and extrapolation from existing relic areas, (2) comparison of grazed with ungrazed areas, (3) evaluation and interpretation of research data on plant communities, and (4) review of historical accounts and botanical literature (Pendelton, 1989). Such a process can be a difficult and time-consuming endeavor, however.

Alternative Approaches to Site Classification

Because of the problems with basing site distinctions on differences in climax vegetation, several general suggestions for change have been made by various groups and individuals.

Bentley and Talbot (1951) classified different kinds of rangelands on the basis of the amount of vegetation each type of rangeland could produce, without regard to species, on California grasslands where exotic grasses that have been introduced by humans dominate. On these sites, estimation of the botanical composition of the climax vegetation has little relevance to current management since the ability to return to climax vegetation has been eliminated by the introduction of exotic species.

The Range Inventory Standardization Committee of the Society for Range Management (1983) recommended the term "ecological site" for the basic unit of rangeland classification. That committee defined ecological site as a kind of land that differs from other kinds of land in its potential natural community and physical site characteristics and, therefore, also differs in its ability to produce vegetation and its response to management.

Recently, the Society for Range Management recommended that land types that differ significantly in their ability to produce vegetation (either kind or amount) should be called *ecological sites* and should be defined as a kind of land with physical characteristics that differ from those of other kinds of land in their ability to produce distinctive kinds and amounts of vegetation and in their response to management. The difference in this definition from that of the earlier committee (Society for Range Management, Range Inventory Standardization Committee, 1983) is the omission of a reference to a potential natural community (USFS and BLM). Justification for using *ecological site* (BLM) rather than *range site* (SCS) is based on the reasoning that site classification is not necessarily oriented to any particular land use or land type.

Considerable evidence exists in the ecological and agronomic literature for the classification of rangelands into sites on the basis of relationships among climate, soil, and vegetation. In nature, these boundaries are arbitrary, meaning that all site descriptions are approximate but useful classifications that can be used to organize management knowledge and research results.

Because site classification is an arbitrary division of the landscape and because individual sites exist within a patchwork of other sites, some considerations of influences at the landscape level may be desirable for some purposes. Australian researchers have developed land systems models that describe sites that occur together on rangelands (Mabbut, 1968). Similar to soil associations, these land systems models relate adjacent sites that may or may not have common characteristics. Analysis of the ways in which sites are associated and their interactions within a landscape could improve understanding of ecosystem processes at the multiple-site level, where most management implementation occurs. For example, landscape position may be important when sites that receive runoff, such as riparian zones, are being eroded because of vegetation conditions on other parts of the watershed.

SITE CLASSIFICATION FOR ASSESSING RANGELAND HEALTH

SCS, USFS, and BLM should adopt common site classifications for the purpose of coordinating rangeland health inventories and monitoring efforts.
Although methods of site classification between the agencies are similar in concept, differences in the definitions of climax plant communities (SCS) and potential natural communities (USFS and BLM) make comparisons of assessments on different administrative units difficult. It is important that all three agencies (BLM, SCS, and USFS) use similar classification methods, so that the results of management, research, and assessments can be compared across administrative boundaries.

To limit interpretation conflicts that could arise from the proliferation of various classification schemes for various purposes, common site classifications should be soil based and should provide general information on vegetation production and life-form dynamics. They should also describe responses to disturbances such as fire, grazing, and drought. For the purposes of rangeland health assessments, it would be useful to add descriptors of soil surface, nutrient cycles, energy flows, and recovery mechanism attributes to the current criteria for describing range sites (SCS) or ecological types (USFS).

Range Condition: Ecological Status

Traditionally, range scientists have defined the term "range condition" to mean "the state of range health" (Society for Range Management, 1989:2), and federal management agencies have, in fact, used the term in this manner (see, for example, U.S. Department of the Interior, Bureau of

Land Management [1984], and U.S. Department of Agriculture, Soil Conservation Service [1989a]). The development of the current methods for evaluating the ecological state of rangelands on the basis of the departure from climax vegetation and the succession-retrogression model of rangeland change can be viewed as the first approximation of rangeland health. There were and are reasons to consider rangelands that contain climax vegetation healthy, as defined by the committee. The process of ecosystem development—that is, succession—was thought to culminate in maximum stability, productivity, diversity, and other presumed desirable qualities (Stoddart et al., 1975). Communities in the early stages of succession were thought to be characterized by less complex energy flows and more open nutrient cycles and to be more vulnerable to invasion by exotic species (Odum, 1969). Assessments of range condition (SCS) and ecological status (USFS and BLM) have produced a wealth of useful data and research that has provided the underpinning for efforts to manage the impact of grazing on both federal and nonfederal rangelands.

Range condition (SCS) and ecological status (USFS and BLM) ratings, however, are not sufficient measures of rangeland health, as defined in this report. The committee identified three problems with range condition (SCS) and ecological status (USFS and BLM) that limit the utility of these methods as measures of rangeland health: (1) use of climax plant community (SCS) or potential natural community (USFS and BLM) composition as standards, (2) the difficulty in identifying thresholds of change, and (3) the reliance on changes in plant composition and production as the sole indicator of change in the ecological state of rangelands.

STANDARDS FOR ACCEPTABLE CONDITIONS

The current methods of assessing range condition (SCS) or ecological status (USFS and BLM) establish a benchmark plant community against which current plant composition and production are compared. Condition and status ratings are a measure of how closely the current vegetation resembles the defined benchmark plant community.

LINKING SUCCESSIONAL STAGE TO CONDITION AND STATUS RATINGS   The linking of successional stages to range condition (SCS) classes has confused the interpretation of the results of range condition (SCS) and ecological status (USFS and BLM) surveys. Smith (1989) has observed that acceptance of the view that succession is ecosystem development that culminates in maximum stability, productivity, diversity, and other presumed desirable qualities "really leaves one with little choice but to manage for near climax, or admit that the goal is a second rate, degenerated ecosystem" (Smith, 1986:120). The public, the U.S. Congress, and environmental interests are understandably concerned with reports that 36 and 16 percent of lands managed by BLM are in fair or poor condition, respectively. Concern that poor or fair condition does indicate rangeland degradation is reinforced when decreases in the amount of rangelands in poor or fair condition are reported as evidence of agency success in meeting mandates to improve rangelands (see U.S. Department of the Interior, Bureau of Land Management [1990], for example). Rangeland managers and livestock producers respond that fair and poor conditions do not necessarily indicate a problem or the need for changes in management practices on a particular site; they only indicate that rangeland's stage of succession. BLM and USFS have eliminated the terms excellent, good, fair, and poor and have adopted terminology that reflects successional stages. The problem that remains, however, is in determining whether there is a cause for concern about any one of the successional stages that a rangeland may be in.

The relationship between successional stages and the stability of a site's soil and the integrity of its ecological processes—that is, its health—is uncertain. Spence (1938) noted that the soil, water, and productiveness of a rangeland are conserved when it contains its climax vegetation but that rangelands in earlier stages of succession can also conserve these values. Spence also noted that species that are not part of the climax vegetation can also conserve the soil, water, and productiveness of rangelands.

Lauenroth (1985) concluded that the similarity of a rangelands' vegetation to climax vegetation does not necessarily indicate where site degradation is occurring or where it might occur. Different combinations of plants with the same degree of similarity to an established benchmark community may differ in the effectiveness with which they protect the site from accelerated soil erosion. Risser (1989) noted that a rangeland's vegetative composition may not reflect the erosional status of the soil mantle, and Wilson (1989) reported that there is no clear relationship between changes in plant composition and soil erosion on rangelands studied in Australia. Smith (1989) also noted climax vegetation is not the only type of vegetation that furnishes adequate soil protection. Other investigators have noted that seedlings of exotic species, such as crested wheatgrass (*Agropyron cristatum*), that are completely dissimilar to the climax vegetation can offer adequate protection from erosion (Dormaar et al., 1978).

The committee inspected a site near Reno, Nevada, that illustrates the difficulty in clearly relating the degree of similarity to an established benchmark plant composition to the degree to which the soil and ecological processes are being conserved. The existing plant composition was dominated by desert needlegrass (*Stipa speciosa*) and Wyoming big sagebrush (*Artemisia tridentata* ssp. *wyomingensis*) but included a number of other species such as Indian rice grass (*Oryzopsis hymenoides*) and Utah juniper (*Juniperus osteosperma*). The percent species composition was similar enough to the climax plant community (SCS) so that the rangeland could be rated in the excellent condition class on the basis of the rangeland's plant composition alone. The range condition (SCS) of the site was reduced to good, however, because of serious wind erosion. Wind erosion on the site was severe enough that many plants were being damaged by abrasion, and some were being buried by drifting soil. It appeared that the amount of plant cover on the site had declined, exposing the soil to wind erosion. The decline in plant cover, however, appeared to have been evenly distributed among all species present on the site, so that the percent composition remained the same as that which would be expected in the climax plant community (SCS). Similarity to a benchmark plant composition was not, in this case, a sensitive indicator of other changes that were occurring on the site.

The similarity of current plant composition and biomass production to that of a climax plant community (SCS) or potential natural community (USFS and BLM) should not be used as the primary standard of rangeland health.

The degree of similarity to the plant composition and annual biomass production of a climax plant community (SCS) or potential natural community (USFS and BLM) alone is not a sufficient measure of rangeland health. The evaluation of rangeland health will require additional and different criteria and indicators.

DIFFICULTIES IN COMPARING DIFFERENT SITES   The climax plant community (SCS) and the potential natural community (USFS and BLM) is different for each site. Similarly, the plant composition and biomass production of successional stages also may differ between sites. Since the benchmark against which current plant composition is compared is different for different sites, comparing the results of condition or status ratings between sites is difficult. It is possible, for example, that an early successional stage on one site may closely resemble a later successional stage on a different site. Plant composition and production may be very similar between the two plant communities on the two sites, but they may receive different condition or status ratings since the benchmarks against which they are being compared differ.

Laycock (1989) cited an example of such a problem in Idaho:

The *Artemisia tridentata* ssp. *wyomingensis/Poa secunda* [Wyoming big sagebrush/sandberg bluegrass] habitat type occurs in areas in western Idaho where precipitation is less than 18 cm annually (Hironaka et al., 1983). The dominant plant is the Wyoming big sagebrush with an understory of sandberg bluegrass and scattered other species. This [Wyoming big sagebrush/sandberg bluegrass] vegetation mix is identical to a rather severely degraded ("poor condition") stage of the *Artemisia tridentata wyomingensis/Stipa thurberiana* [Wyoming big sagebrush/Thurbers needlegrass habitat] type that occurs over a wide area of southern Idaho on better sites and higher precipitation where grazing has removed the [needlegrass] and other desirable species. It may also resemble other sagebrush habitat types in lower stages of condition. (Laycock, 1989:5)

The sites described by Laycock had essentially the same plant composition and production, but they would have different ratings since the benchmarks against which they were compared differed.

This difference may be important because the relationship between a similarity rating and the conservation of soil and ecological function is not well understood. One site that supports 40 percent of its expected climax plant community (SCS) or potential natural community (USFS and BLM) composition may still have enough soil cover to be protected from water and wind erosion, whereas another site, also supporting 40 percent of its expected climax plant community (SCS) or potential natural community (USFS and BLM) composition, might be eroding at rates that will lead to serious site degradation, it might have a compacted soil that restricts infiltration, or it might have nutrient cycles that are interrupted by a lack of litter cover and lack of incorporation of organic matter into the soil surface.

The use of similarity to a defined benchmark plant community, whether defined as climax plant community (SCS) vegetation or potential natural community (USFS and BLM), also imposes the difficulty of direct comparisons between sites with different benchmark plant communities. The lack of a single, clearly defined standard that does not differ from site to site is a fundamental limitation to the use of current methods for assessing rangeland health. This problem has and continues to confuse the public, the U.S. Congress, livestock producers, and rangeland scientists themselves.

LIMITATIONS OF SUCCESSION-RETROGRESSION MODEL

The successional model that supports the current range condition (SCS) and ecological status (USFS and BLM) methods postulates that succession toward the climax plant community (SCS) or potential natural community (USFS and BLM) is a process that moves through recognizable and predictable stages. The effects of grazing, drought, and other disturbances are also thought to produce recognizable and predictable changes in composition toward early stages in succession. The vegetation that is found on any particular rangeland is the product of the equilibrium between the two forces of succession toward the climax plant community and retrogression toward earlier successional stages.

This rangeland succession model assumes that range condition (SCS)—that is, the successional stage of a particular rangeland—will change back and forth along the successional gradient characteristic of that site in response to changes in management. The main tool of rangeland managers is to adjust the stocking rate, species of grazing animal, the duration of grazing, and the season that the rangeland is used to achieve an equilibrium between the opposing forces of succession and retrogression. Once this equilibrium is achieved, it will tend to remain stable. The manipulation of livestock grazing allows rangeland managers to maintain a particular plant composition at some point along the successional gradient characteristic of that site. The choice of that point along that gradient that should be maintained depends on whether the goal of management is to maximize or optimize the production of livestock, wildlife, or some other product or value.

DOES SUCCESSION OCCUR ON ALL RANGELANDS? Current ecological research has questioned whether the concept of well-defined, predictable, and reversible changes along a successional gradient holds for all or the majority of rangelands.

Ecologists working in rangeland ecosystems have developed theories that allow for multiple equilibria and for transitions between alternative vegetational states that are not easily reversible (Friedel, 1991; Westoby et al., 1989). These theories, which were discussed in more detail in Chapter 2, also allow for the existence of transitional states that represent the process of change from one state of equilibrium to another. The specific outcome of the change will depend on events that occur while the rangeland is in that transitional state rather than on a predictable succession from one state to another. Investigators have attempted to describe the mechanisms that produce such complex dynamics on rangelands. In some cases, the random occurrence of fire, drought, or changes in grazing systems have produced changes in rangelands that do not appear to follow a readily discernible successional sequence (Friedel, 1991; Laycock, 1989; Westoby et al., 1989).

Sharp et al. (1990), for example, reported 40 years of data from a salt desert shrub rangeland in south-central Idaho that illustrate this phenomenon. Three different plant communities have occurred on this site, which has not been grazed since 1945. During periods of normal precipitation and no buildup of a naturally occurring scale insect (*Orthisae* sp.), the site is dominated by shadscale (*Atriplex confertifolia*) in association with bottlebrush squirreltail (*Sitanion hystrix*), sandberg bluegrass (*Poa secunda*), globe mallow (*Sphaeralcea coccinea*), and other plants. Following a cyclic outbreak of the scale insect, most of the shadscale dies or is reduced in size and vigor. If precipitation is normal in the years following the outbreak, the site becomes dominated by bottlebrush squirreltail, production is high, and shadscale may eventually recover. If precipitation is below average following the insect outbreak, several species codominate the site. The species that becomes more important depends on the timing of precipitation events. The transition from one complex to another appears to be due more to the vegetation's adaptation to episodic events than to a linear successional development. It could be argued, however, that the desert sites described simply have not had time to recover from historic overgrazing and to reestablish the potential natural community (USFS and BLM).

In any case, the kinds of vegetation dynamics described above are difficult to incorporate into existing succession-retrogression models of rangeland development.

Risser (1989) summarized the questions about the succession-retrogression model raised by ecologists as including some of the following ideas. Biological communities may go through different pathways yet reach a similar climax or terminal state. Depending on disturbances during succession, the system may proceed through a new set of seral stages. The initial conditions at the beginning of a successional sequence can cause quite different outcomes even on apparently equivalent sites. The outcome of successional sequences is determined by the characteristics of the interacting plant and animal populations and the present and preceding environment, not by predetermined organismic controls. Certain ecosystem-level characteristics, such as the ability to absorb or release nutrients, are characteristic functions of systems in early and late stages of succession. The terminal stage may not be the most productive, stable, or diverse community.

Risser concluded that it is important to recognize that a simple linear successional sequence is not always an adequate representation of the conditions observed in the field (Connell and Slatyer, 1977; Drury and Nisbet, 1973; Gutierrez and Fey, 1975; McIntosh, 1980; Odum, 1969 [as cited by Risser, 1989]).

Traditional successional theory implies that a site that has retrogressed can recover if the process is reversed. This is not possible or is very slow, however, if severe soil erosion, invasion of a new and very dominant species, or change from a fire-dependent to a fire-safe plant community has resulted in near-permanent changes in the abiotic or biotic community.

The succession-retrogression model of how rangelands develop and change, which is the foundation of current rangeland classification and assessment methods, should be modified or new models developed to assist in assessing whether rangelands are approaching thresholds of change.

Current range condition (SCS) and ecological status (USFS and BLM) ratings are founded on the concept of well-defined, predictable, and reversible changes along a successional gradient that holds for all or the majority of rangelands. An evaluation of rangeland health requires consideration of additional processes of ecosystem change and the reversibility of those changes. Several alternative models of rangeland change have been proposed, but no single model has received widespread acceptance. An accelerated effort by ecologists is needed to develop and test models of rangeland change that will assist in identifying rangelands that are approaching thresholds of change.

MULTIPLE INDICATORS ARE NEEDED

Current range condition (SCS) or ecological status (USFS and BLM) ratings rely nearly exclusively on measurements of plant composition and annual biomass production. Problems such as soil erosion, disruption of nutrient cycling, or other ecological attributes of rangelands are not primary considerations in assessing range condition (SCS) or ecological status (USFS and BLM). Changes in other important attributes of an ecosystem may not be detected by measuring the plant composition and production alone.

The difficulty of assessing ecosystems using only one index has long been recognized. Ellison (1949) wrote that the soil, plants, animals, topography, and climate develop together and are knit together into an integrated whole. He also noted that vegetative and soil trends do not always parallel each other and that they may be widely divergent on eroding rangelands. He suggested that "if the rangeland manager is impressed by evidence of change in vegetation and not by evidence of soil erosion, he may be led astray" (Ellison, 1949:794).

The effectiveness of vegetation in protecting soil is more a function of effective soil cover than plant composition, since effective soil cover is more closely tied to the type and pattern of cover than it is to plant composition. Slight changes in plant populations and litter may produce accelerated erosion on steep, erodible soils, whereas complete changes in species and life-forms, such as in artificially established annual grasslands, may result in a vegetation type that still provides good soil protection. Erosion hazard on some soils may be quite insensitive to changes in vegetation type if the site is very flat or very rocky, whereas on steep clay slopes the kind or amount of vegetation may be critical (Smith, 1989).

Loss of minor species may not be indicated by a change in range condition (SCS) or ecological status (USFS and BLM) rating if these species make up a small percentage of the plant composition and annual biomass production of the climax plant community (SCS) or potential natural community (USFS and BLM). The loss of minor species, however, may indicate change in nutrient cycles caused by reduced diversity in rooting depth or changes in energy flow because of a reduced period during which the remaining plants photosynthesize.

Tueller (1973) described a process of site degradation that began with the loss of plant vigor and seed production and that led to the death of individual plants and a reduction in litter cover and plant density. These changes caused changes in plant cover, distribution, and potential for reproduction. Total biomass production or the annual production of individual species was reduced. Further deterioration led to reduced litter accumulation, the formation of soil crusts that retarded germination, and altered plant growth forms. Reduction in soil cover and litter led to soil erosion and the disruption of nutrient cycles. Eventually, the site potential was seriously impaired.

The process of site degradation described by Tueller (1973) is driven by a complex of interacting factors, with no single factor predominating. Changes in species composition, plant density or frequency, distribution and cover of litter, soil erosion, total biomass production, plant vigor, and seedling recruitment, among other factors, are all apparent at different points in the degradation process. No single factor alone can completely describe such a process. Tueller for instance, listed 16 separate factors that can serve as useful indicators of site degradation (Tueller, 1973).

The problem is not that plant composition and biomass production are unimportant attributes of rangeland ecosystems; rather, the problem is that they are typically the only attributes measured. A system that used only plant density, erosion, litter cover, seedling density, or compaction as a single measure of range condition (SCS) or ecological status (USFS and BLM) would have the same problem.

The evaluation of rangeland health will require analysis of attributes in addition to plant composition. A comprehensive evaluation of rangelands should be based on the preponderance of evidence derived from sampling multiple attributes related to ecological function and soil stability. Range condition (SCS) or ecological status (USFS and BLM) ratings based primarily on changes in plant composition and annual biomass production alone are not sufficient measures of rangeland health.

Plant composition and biomass production will usually change if soil erosion accelerates, the soil continues to compact, or seedlings fail to become established. However, there may be significant lags between the onset of rangeland degradation and a change in any single indicator of rangeland health. Any single measure will have different sensitivities to different stresses. The measurement of multiple attributes increases the probability that rangeland degradation will be detected early enough for corrective measures to be taken.

Site Potential and Resource Values

In current theory and practice, the rangeland site classification and the definition of site potential are nearly synonymous. The kinds and amounts of vegetation produced in the climax plant community (SCS) or potential natural community (USFS and BLM) are considered the best measure of the potential productivity of a site. Overgrazing, accelerated erosion, or other influences that result in loss of the capacity to produce the plant composition and annual biomass production characteristic of the climax plant community (SCS) or potential natural community (USFS and BLM) are thought to have caused a loss in site potential.

DESIRED PLANT COMPOSITION

Many observers have argued that the plant composition and production desired on a site should depend on how the site is to be used. In its *National Range Handbook* (U.S. Department of Agriculture, Soil Conservation Service, 1976), SCS states that although the climax plant community (SCS) describes the site potential of a particular site, the goal of management is not necessarily always to achieve climax plant community (SCS) composition and production. Other seral (successional) stages may be better for particular uses. Risser (1989) stated that the climax vegetation for a given site may not by the most productive or desirable type of vegetation for livestock forage production and that climax vegetation may not be the most appropriate goal of rangeland management if the management objectives include multiple uses or values. Such multiple uses or values may include wildlife, water quality or quantity, recreation, and livestock grazing.

RESOURCE VALUE RATING

Wilson (1989) suggested the need to first define the land use objectives for rangelands. This should be followed by a description of the vegetative structure that will maximize those objectives. The Society for Range Management's Range Inventory Standardization Committee similarly recommended that a system of resource value ratings be used as a measure of the value of vegetation or other features for a particular use (Society for Range Management, Range Inventory Standardization Committee, 1983). The resource value rating would be based on the particular species present, growth forms and foliage types, or other criteria. Each use may have a separate resource value rating. The resource value rating was to be a measure of the suitability or usefulness of the vegetation of an ecological site (BLM) for a specific use.

Most recently, the Society for Range Management's Task Group on Unity in Concepts and Terminology recommended that management objectives should be defined in terms of a desired plant community for each ecological site (BLM). The desired plant community should be defined as follows: "of the several plant communities that may occupy a site, the one that has been identified through a management plan to best meet the plan's objectives for the site" (Society for Range Management, Task Group on Unity in Concepts and Terminology, 1991:10).

USFS has proposed adoption of a system of resource value ratings. Resource value ratings require that goals be set for the vegetation needed to meet a particular use. Those goals will be a desired plant community. Because goals will vary from location to location, different desired plant communities may be described for different locations on the same site. Management would be designed so that the plant composition could be changed to reflect that described for the desired plant community. This may or may not represent a change toward the climax plant community (SCS) or potential natural community (USFS and BLM).

DISTINCTION BETWEEN RANGELAND HEALTH, SITE POTENTIAL, AND RESOURCE VALUES

Rangeland health assessments should be separate and independent of assessments for determining the proper use of particular rangelands.

It is essential that there be a clear understanding of the distinction between rangeland health, as defined in this report, and site potential or resource value ratings. Rangeland health, as used in this report, is intended as a minimum ecological standard, independent of the rangeland's use and how it is managed. The particular mix of commodities and values produced by a rangeland depends on how it is used and managed. If rangeland health is conserved, then the capacity of the site to produce different mixes of commodities and values is conserved. The determination of which uses and management practices are appropriate may require the evaluation of data different from those used to evaluate rangeland health. No single index will meet all the needs of rangeland inventory, classification, and management.

Rangeland health is a measure of the integrity of the soil and the ecological processes of a rangeland. Loss of rangeland health causes a loss of capacity to produce resources and satisfy values. Rangeland health is not, however, an estimate of the kinds or amounts of resources that a rangeland produces, nor is it an evaluation of the different uses of a site. Two rangelands may have the potential to produce different commodities and values, but both can be equally healthy if the integrity of the soil and ecological function are conserved.

The protection of rangeland health should serve as the minimum standard for management. If rangeland health is sustained then decisions about the appropriate plant community composition and production can be made depending on the desired rangeland use. Most important, the conservation of rangeland health preserves the option to change the use and management of a site as the desired resources and values change.

Apparent Trend

A one-time measure of most rangeland characteristics is only that—a picture of the situation at the time of measurement. Without a previous measurement with which the current measurement can be compared, the range manager's ability to interpret whether the management program is succeeding or failing is limited. Personnel and budget constraints and inconsistencies in the indicators measured at different times, however, have limited the number of sites where trend can be determined from comparable data collected at different points in time.

To compensate for this problem, range managers have attempted to determine apparent trend; that is, they evaluate site characteristics that indicate whether an area is improving or deteriorating. Factors such as accelerated erosion, for example, have been used to indicate a downward trend. If the soil is eroding at an accelerated rate, then the productive capacity of the site is probably being lost. The accumulation of litter is viewed as a sign of an improving rangeland because it is a sign that the amount of plant material needed to protect a site from erosion is increasing. The presence of seedlings of desirable plants was also interpreted as an upward apparent trend because those seedlings indicated that the plant composition was evolving toward the climax plant community (SCS) or potential natural community (USFS and BLM). The presence of seedlings of undesirable plants represents a downward trend. Plant vigor has also been used to judge apparent trend. Vigorous dominant plants in a climax plant community (SCS) or potential natural community (USFS and BLM) indicate an upward apparent trend, whereas the presence of weak, deformed plants indicates a downward trend.

4 Criteria and Indicators of Rangeland Health

Rangeland ecosystems are continually responding to temporary changes in the physical and biotic environments. A system that assesses rangeland health must be able to distinguish between changes that result in the crossing of a threshold from those that are temporary because of normal fluctuations in physical or biotic factors. Some of these changes, that is, threshold shifts, may be difficult to reverse, but they do not necessarily entail a loss of the capacity to produce commodities and satisfy values. The process of rangeland degradation is complex and involves the interaction of changes in the physical, chemical, and biological properties of soils, as well as changes in plant vigor, species composition, litter accumulation and distribution, seed germination and seedling recruitment, total biomass production, and other ecological functions. Tueller (1973) reviewed the process of rangeland degradation and suggested 16 factors that operate at different stages of the degradation process. The process of rangeland improvement is just as complex. Multiple criteria and indicators will be needed to assess whether rangelands are healthy, at risk, or unhealthy.

It is also clear that the evaluation of rangeland health is a judgment, not a measurement. Rangeland health, like range condition (SCS) or ecological status (USFS and BLM), is not a physical characteristic of rangelands that can be measured directly. The indicators of rangeland health, range condition (SCS), or ecological status (USFS and BLM) can, however, be measured. The evaluation of rangeland health will require judgments on the significance and meaning of the indicators that are measured. Evaluation of the preponderance of evidence from the evaluation of multiple indicators will be required for a meaningful assessment of rangeland health.

The determination of whether a rangeland is healthy, at risk, or unhealthy should be based on the evaluation of three criteria: degree of soil stability and

Desert needlegrass (*Stipa speciosa*)

97

watershed function, integrity of nutrient cycles and energy flow, and presence of functioning recovery mechanisms.
The process of rangeland change is complex, and multiple criteria should be used to determine whether rangelands are healthy, at risk, or unhealthy. No single criterion alone will be a sufficient basis for this determination. The committee recommends a three-phase approach for assessing rangeland health. Phase 1 is an evaluation of soil stability and watershed function. Phase 2 is an evaluation of the functioning of nutrient cycles and energy flows. Phase 3 is an evaluation of the probability that recovery mechanisms will occur on the rangeland being assessed.

SOIL STABILITY AND WATERSHED FUNCTION

The physical, chemical, and biological processes that occur in rangeland soils supply plants with nutrients and water. Microorganisms in the soil break down plant litter, releasing nitrogen, phosphorus, and other nutrients essential to plant growth. The texture, structure, and porosity of soil determine how much rain is captured and how much runs off during a storm. Soils are storehouses of water and nutrients for plants to draw on when they need them. The soil is a living system that is inextricably linked to nutrient cycles, energy flows, and other ecological processes of rangeland ecosystems.

Soil Degradation

There are three principal processes involved in soil degradation: physical, chemical, and biological. These processes are closely linked, and modification of one unavoidably alters the others. Physical degradation results in the deterioration of the physical properties of soils through compaction, wind or water erosion, deposition of sediments, and loss of soil structure. Biological degradation occurs when there is a reduction in the organic matter content of the soil, a decline in the amount of carbon stored as biomass, and a depression in the activity and diversity of the organisms living in the soil. Chemical degradation includes nutrient depletion, shifts toward extremes in the pH of the soil, increases in salt concentration, and contamination by toxic substances such as heavy metals (Lal and Stewart, 1990). Summaries of these phenomena and interactions can be found in basic soils texts (for example, Brady [1990], Foth [1990], Miller and Donahue [1990], and Singer and Munns [1987]).

SOIL EROSION BY WIND AND WATER

Soil erosion by wind and water is a major factor in the process of soil degradation on rangelands and has been recognized as such for a long time. It affects the physical, chemical, and biological properties of soils. Lincoln Ellison wrote: "we know that range condition ceases to be satisfactory when accelerated soil erosion sets in, when destructive processes clearly exceed constructive processes. Hence a basic criterion of range condition is degree of soil erosion, and a minimal requirement for satisfactory condition is normal soil stability" (Ellison, 1949:790).

Ellison considered the soil to be an index of the extent to which soil, plants, animals, and climate are knit together into an integrated whole. To Ellison, the presence of a well-developed soil was evidence of successful integration of climate and topography, vegetation, and animal life over a long time period. Ellison indicated, however, that accelerated soil erosion is evidence of disintegration and of a relatively recent change in the relationship between components of the range complex that were formerly well integrated. It is also an indication that a change of drastic proportions, over and above the normal amplitude of environmental stress, has taken place.

EFFECTS OF SOIL DEGRADATION

Soil degradation has profound effects on rangeland ecosystems. Smith (1989) concluded that site deterioration occurs mainly through deterioration of the soil's capacity to capture and store water, loss of the ability of the soil to supply nutrients, or the accumulation of salts or other toxic substances in the soil. Erosion and deposition of eroded sediments are, according to Smith (1989), major processes of site degradation; but degradation of the soil's structure, losses of nutrients to the atmosphere by gasification, movement of dissolved nutrients beyond the reach of plant roots by water percolation through the soil, or changes in the depths to water tables also cause rangeland degradation.

Wilson and Tupper (1982) suggested that there are four classes of rangelands based on whether the soil is stable or unstable and whether vegetative productivity is good or diminished. Friedel (1991) wrote that site deterioration is best indicated by irreversible changes in the soil, and concluded that assessment of the soil surface is a critical element in the identification of thresholds of change on rangelands. Similarly, Risser (1989) and Wilson (1989) emphasized the importance of soil erosion in rangeland assessments and recommended that soil criteria be incorporated into current assessments of range condition (SCS) and ecological status (USFS and BLM). Most recently, the Society for Range Management's Task Group on Unity in Concepts and Terminology (1991) recommended that the effectiveness of present vegetation in protecting a site against accelerated erosion by water should be assessed independent of the use of the site and that any site determined to be suffering accelerated erosion should be considered in unsatisfactory condition.

OTHER EFFECTS OF SOIL DEGRADATION

Soil degradation affects not only soil attributes but can also degrade other ecological processes. Loss of organic matter in the soil reduces nutrient stores and interrupts nutrient cycles. Accelerated soil erosion reduces the total organic matter and total nitrogen contents of soils and the capacity of rangeland soils to hold moisture (Croft et al., 1943). The formation of soil crusts and the development of erosion pavement (a hard, impermeable soil surface caused by erosion) can impede germination and growth of seedlings (Blaisdell and Holmgren, 1984; Troeh et al., 1991). Reduced water infiltration and water storage can reduce total vegetative biomass production and can result in shifts in species composition (Archer, 1989).

SOIL STABILITY AND THE ENVIRONMENT

Given the importance of soil stability, it is important to recognize that rangelands are often located in arid or other extreme environments where the processes of soil development are slow or impeded (Hugie et al., 1964; Passey et al., 1982; Wooldridge, 1963). Destructive processes such as wind and water erosion can easily exceed constructive process such as the accumulation of soil organic matter. Naturally destructive processes are thus highly probable on many rangelands (Friedel, 1991).

Managers of rangelands must minimize the consequences of processes that destroy the soil if rangeland health is to be conserved. There are some arid, steeply sloping, or other sites in extreme environments, however, where destructive processes dominate and render even good management inconsequential. On such sites soil instability is manifest in the lack of soil horizons (the presence of a soil horizon is a characteristic of developed soil), little organic matter accumulation, and limited development of plant communities. Such sites have been unstable for millennia and will continue to be unstable long into the future. They are the result of processes that occur over geological time scales (tens of thousands of years) and can be considered naturally unhealthy or at risk. Management options that conserve or encourage the development of rangeland health on such sites are limited (see discussions by Retzer [1974] and Birkeland [1984]). These sites will require careful management to ensure improper use does not accelerate the destructive processes already operating on the site.

Rangelands in less severe environments are exposed to varying degrees of soil degradation. Animal hooves may cause mild compaction of the soil in some locations. Prolonged wheeled vehicular traffic causes significant compaction on most soils. Abusive overgrazing not only may

Overgrazing of livestock is an important factor in deterioration. Wind erosion takes a heavy toll on soil in this condition. Credit: U.S. Department of Agriculture.

cause compaction but may remove plant cover and open the way for destructive erosion as well (Dormaar and Williams, 1990). Little research has been done to determine whether grazing causes loss of soil organic matter, but some studies have shown that wild and domestic grazing animals can influence soil processes (Pastor et al., 1988; Whicker and Detling, 1988). However, any factor that severely reduces soil cover opens the way for removal of organic matter via erosional events.

EFFECTS OF SOIL DEGRADATION ON WATERSHEDS

Soil degradation directly influences watershed function on rangelands. Precipitation that falls on rangelands ultimately infiltrates into the soil, evaporates, or becomes part of the overland flow and runoff to surface water (Branson et al., 1981). Figure 4-1 describes the pathways taken by rainfall on rangelands. The proportion of precipitation captured de-

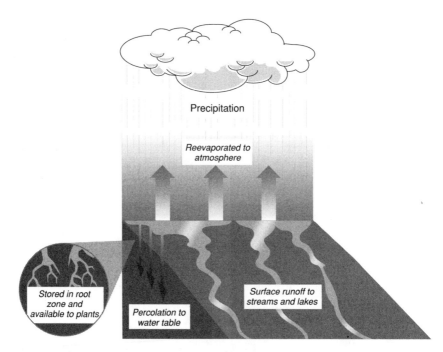

FIGURE 4-1   The pathways taken by rain falling on vegetated land. Source: Derived from R. Brewer. 1989. The Science of Ecology. Philadelphia: Saunders College Publishing.

pends on the energy, time of year, and total volume of precipitation during a particular rainfall event and by the properties of the soil on which it falls (Branson et al., 1981). Texture, structure, moisture content, vegetative or litter cover, and organic matter content are the most important properties of soil that influence infiltration (Satterlund, 1972).

Degradation of watershed function has direct effects on rangelands. Grazing, gathering of fuelwood and harvesting of woody and herbaceous vegetation, and fires all may have dramatic and direct effects on the infiltration of water into the soil. Standing vegetation, litter and duff (partly decayed organic matter), and organic matter incorporated into the soil all improve the ability of water to infiltrate the soil through various processes (Figure 4-2). A site that has lost its vegetation, litter, and ultimately, the organic matter incorporated into its soil tends to seal its surface and encourage water to run off as overland flow rather than be absorbed into the soil profile. As less water infiltrates the soil, less plant growth is possible, and therefore, soil cover and the amount of organic matter in the soil are reduced, leading to less infiltration. This process can lead to a self-reinforcing cycle of watershed degradation as increased volumes of runoff accelerate soil erosion, which, in turn, reduces the infiltration of water into the soil and increases runoff (Shaxon et al., 1989). Loss of the nutrients that are attached to soil particles can lead to further degradation through the loss of soil fertility (Logan, 1990).

Soil degradation, primarily through accelerated wind and water erosion, causes the direct and often irreversible loss of rangeland health. Soil degradation not only damages the soil itself but also disrupts nutrient cycling, seed germination, seedling development, and other ecological processes that are central to rangeland health. In addition, soil degradation can degrade watersheds, leading to the further loss of rangeland health as well as off-site damages to water quality.

Criteria and Indicators of Soil Stability and Watershed Function

The criteria and indicators that are selected to assess soil stability and watershed function must relate to two fundamental processes: soil erosion by wind and water and infiltration or capture of precipitation. These

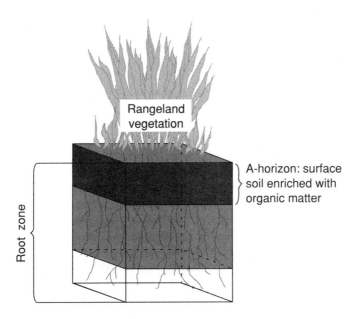

FIGURE 4-2 Diagram of a soil profile. Rangeland vegetation helps to create a layer of surface soil rich in organic matter, which serves as a basis for an extensive root system. In such a profile, water is better able to infiltrate the soil, limiting the effects of erosion.

two processes interact with each other, and measurable indicators of their activities can provide data that can be used to evaluate the stability of soils and interpret whether watershed functions are adequate.

The degree of soil movement by wind and water should be the criterion used to assess soil stability and watershed function; evaluation of soil movement should be based on multiple indicators of the condition of the soil surface.

Soil erosion by wind and water is the most important form of soil degradation on rangelands, and erosion damage is often irreversible. The degree of soil movement by wind and water should be fundamental to the assessment of rangeland health.

The use of soil surface characteristics such as rills and gullies, scours (bare soil caused by the scouring action of wind or water) and dunes, litter cover, flow patterns, and pedestaling (erosion of soil from around the base of a plant such that it appears to be on a pedestal) have long been used as indicators of soil stability. Table 4-1, which was developed by the Bureau of Land Management, is an example of the use of soil surface characteristics to assess soil stability.

Table 4-2 lists the general surface soil characteristics that are useful indicators of the degree of the soil stability and watershed function. The development of rills and gullies, the degree of pedestaling, evidence of scouring by wind or sheet erosion (erosion caused by thin sheets of water running off unprotected soil), and deposition of eroded material in fans or dunes are all evidence of soil erosion and runoff. The presence and distribution of an organically enriched A-horizon (the uppermost layer of soil in which organic matter is deposited and the decay process occurs) is evidence of the balance between soil development and degradation. All of these indicators have been used in the past, and many are currently used by federal agencies to evaluate the stability of rangeland soils.

Estimates of erosion rates on rangelands have been impeded by the lack of models that are well adapted to rangelands. New models that will better simulate rangeland conditions are being developed. Estimates of erosion rates by wind or water alone, however, will not be sufficient unless researchers can determine erosion rates that are acceptable for maintaining rangeland health. In addition, estimation of erosion rates does not necessarily yield estimates of the proportion of precipitation that is effectively captured—an important criterion for the evaluation of watershed function. Development of soil crusts, for example, may simultaneously reduce soil loss and increase runoff.

Development of predictive models that estimate rates of soil loss and infiltration, coupled with the establishment of natural rates of soil loss, could help to quantify the evaluation of soil stability and watershed function as part of rangeland assessments. Soil surface characteristics will continue to provide the best available information with which to evaluate soil stability.

THE A-HORIZON

The A-horizon is the soil layer where organic matter from plant litter, animal manures and other sources begins to decompose and becomes incorporated into the soil. The condition of this organically enriched A-horizon has important effects on soil stability, nutrient cycling, energy flows, and recovery mechanisms. The condition of the A-horizon, for example, greatly influences how rapidly water can infiltrate into the soil or safely be stored at the surface of the soil. The nutrients released through the biological activity that occurs in the A-horizon are critically important to the productivity of the site.

The A-horizon may not be evenly distributed across the landscape, and it may occur as a thin layer and only in association with living plants in some desert situations. In most rangelands, the presence and distribution of an organically enriched A-horizon can serve as a useful indicator of soil development on a site. The absence or fragmented distribution of an A-horizon indicates that soil is not developing or accumulating on the site. On the other hand, it may indicate that soil formation occurs only in relation to a few prominent grasses or shrubs.

RILLS AND GULLIES

Rills and gullies provide channels for the rapid flow of water from the site. This water may contain soil and organic matter as well as nutrients in solution. Evidence of the development and activity of rills and gullies is a useful indicator of the degree of soil erosion and water infiltration on a site. The physical features of rills and gullies—including their depths, distributions, or degree or the development of a dendritic pattern (branching), which is evidence of active erosion—can be used as indicators of the seriousness of erosion and runoff.

SHEET AND SCOUR EROSION

Sheet erosion by water and scour erosion by wind can remove or reduce the depth of the A-horizon from large areas of rangelands. Continued erosion by these processes results in the loss of subsoil layers beneath the A-horizon as well. Scour erosion by wind may result in abrasive damage to plants as the soil and organic particles blow across the land surface. Sheet and scour erosion can occur without the development of rills or gullies. The degree of development of patches of bare soil as a

TABLE 4-1 Surface Soil Characteristics of the Bureau of Land Management

Characteristic	Class 1	Class 2	Class 3	Class 4	Class 5
Soil movement	Subsoil exposed over much of area; may have embryonic dunes and wind-scoured depressions	Soil and debris deposited against minor obstructions	Moderate movement of soil is visible and recent; slight terracing	Some movement of soil particles	No visual evidence of movement
Surface rock and/or litter	Very little remaining (use care on low-productivity sites); if present, surface rock or fragments exhibit some movement and accumulation of smaller fragments behind obstacles	Extreme movement is apparent; large and numerous deposits against obstacle; if present, surface rock or fragments exhibit some movement and accumulation of smaller fragments behind obstacles	Moderate movement is apparent and fragments are deposited against obstacles; if present, fragments have a poorly developed distribution pattern	May show slight movement; if present, coarse fragments have a truncated appearance or spotty distribution caused by wind or water	Accumulation in place; if present, the distribution of fragments shows no movement caused by wind or water
Pedestaling	Most rocks and plants are pedestaled and roots are exposed	Rocks and plants on pedestals are generally evident; plant roots are exposed	Small rock and plant pedestals occurring in flow patterns	Slight pedestaling, in flow patterns	No visual evidence of pedestaling

Flow patterns	Flow patterns are numerous and readily noticeable; may have large barren fan deposits	Flow patterns contain silt, sand deposits, and alluvial fans	Well defined, small, and few with intermittent deposits	Deposition of particles may be in evidence	No visual evidence of low patterns
Rills and gullies	May be present at depths of 8 to 15 cm (3 to 6 inches) and at intervals of less than 13 cm (15 inches); sharply incised gullies cover most of the area, and 50 percent are actively eroding	Rills at depths of 1 to 15 cm (0.5 to 6 inches) occur in exposed areas at intervals of 150 cm (5 feet); gullies are numerous and well developed, with active erosion along 10 to 50 percent of their lengths or a few well-developed gullies with active erosion along more than 50 percent of their length	Rills at depths of 1 to 15 cm (0.5 to 6 inches) occur in exposed places at approximately 300-cm (10-foot) intervals; gullies are well developed, with active erosion along less than 10 percent of their length; some vegetation may be present	Some rills in evidence at infrequent intervals of over 300 cm (10 feet); evidence of gullies that show little bed or slope erosion; some vegetation is present on slopes	No visual evidence of rills; may be present in stable condition; vegetation on channel bed and side slopes

SOURCE: Adapted from U.S. Department of the Interior, Bureau of Land Management. 1973. Determination of Erosion Condition Class, Form 7310-12. May. Washington, D.C.: U.S. Department of the Interior.

TABLE 4-2 Criteria and Indicators of Rangeland Health

Phase	Criteria	Indicators
Soil stability and watershed function	Soil movement by wind and water	A-horizon present Rills and gullies Pedestaling Scour or sheet erosion Sedimentation or dunes
Distribution of nutrients and energy	Spatial distribution of nutrients and energy	Distribution of plants Litter distribution and incorporation
	Temporal distribution of nutrients and energy	Rooting depth Photosynthetic period
Recovery mechanisms	Plant demographics	Age class distribution Plant vigor Germination and presence of microsites

result of scouring or sheet erosion gives evidence of the seriousness of sheet or scour erosion on a site.

PEDESTALS

The occurrence of plants or rocks on pedestals means that the soil has eroded away from the base of the plant or rock and it has become slightly elevated above the eroded surface of the soil. The height of the pedestals and the degree of root exposure can serve as indicators of the degree of soil loss.

DEPOSITION OF ERODED MATERIAL

Finally, the accumulation of eroded materials around plants or in small basins; as sediment in alluvial fans, gullies, streams, or lakes; or as dunes is a good indicator of erosion. The distribution and abundance of deposits indicate the degree of soil movement. Deposits can range from small accumulations around the base of plants or other obstructions to large fan-shaped deposits.

Multiple Indicators of Soil Surface Condition Needed

The presence and distribution of the A-horizon, rills, and gullies; areas scoured by wind or water; and pedestals under rocks and plants are all indicators that are observable on the landscape. These and other soil surface indicators have commonly been used to assess the degree and severity of erosion on rangelands. Interpretation and judgment are needed to determine whether the problems are large or small or whether the problem occurs on some or most of an area being studied. No single indicator alone is sufficient for an assessment of soil stability and watershed function. The assessment of rangeland health using the criteria described above and summarized in Table 4-2 must be based on the preponderance of evidence obtained from the site.

The Soil Conservation Service (SCS), the U.S. Forest Service (USFS), and the Bureau of Land Management (BLM) should cooperatively develop and implement a system that evaluates multiple soil surface indicators to assess the degree of soil stability on rangelands. An evaluation of soil stability and watershed function, as determined by the use of measurable indicators of the condition of the soil surface, should become a fundamental component of all inventorying and monitoring programs for federal and nonfederal rangelands.

An example of pedestal formation at an early stage. The soil has eroded away from the base of the plant leaving the plant slightly elevated above the eroded surface of the soil. Credit: Kirk Gadzia and Stephen Williams.

Better Soil Surveys Needed

Soil mapping units need to be identified taxonomically to a level sufficient for site classification. In some cases, this may require identification to the level of the soil series and phase.

More data on soil properties related to soil stability and important ecological processes need to be included in soil surveys. Table 4-3 indicates possible soil characteristics that could be included in soil surveys related to plant growth. Of those characteristics listed in Table 4-3, the soil's organic carbon content and available water-holding capacity may be the most important. The amount of organic carbon in the soil affects the rate at which rainfall is captured, the amount of nutrients stored in the soil, and many other processes important to rangeland ecosystems. Since water is normally the most limiting factor on rangelands, the capacity of the soil to store and supply water to plants determines the mix of plants and total biomass production that can be expected. The percentage of organic carbon (or soil organic matter) should be given, as should the depth to which these accumulations occur.

Many of the data in soil surveys are estimated from previous studies. Actual measurement of data would improve the value of surveys for rangeland health assessments. Such data should include morphological data as well as information on the physical, chemical, and biological properties of soil (see Table 4-3).

Many Rangelands Need Soil Surveys

SCS, USFS, and BLM should accelerate efforts to complete standard soil surveys on all federal and nonfederal rangelands.

Modern soil surveys can be an important source of information for assessing rangeland health. Modern soil surveys, however, have not been done for significant areas of federal and nonfederal rangelands. This lack of basic soil information for rangelands will impede efforts to assess rangelands.

Table 4-4 shows the land areas in 13 western U.S. states for which soil surveys have been completed. The total land area mapped varies between 55 and 100 percent. Not all of these unmapped lands are rangelands; however, these data suggest that for large areas of rangelands, soil survey data are not available for use in rangeland assessments. Soil surveys need to be completed on these lands to expedite assessments.

DISTRIBUTION OF NUTRIENTS AND ENERGY

The ability of plants to grow and develop depends on their capture of nutrients from the soil and energy from the sun. Nutrients stored in the

TABLE 4-3 Characteristics Important for Rangeland Health from Representative Soil Surveys

Survey and Reference	Percentage of the Units Surveyed for which the Following Data Were Obtained										
	Color	pH	Texture	Salinity	AWC	Perme-ability	Exc. Base	Carbonate	CEC	Base Sat.	Organic Carbon
Sheridan, Wyo. (Thorp et al., 1939)	100	28	100	0	0	0	0	0	0	0	0
Elbert, Colo. (Larsen et al., 1966)	100	100	100	100	100	100	0	0	0	0	0
Stephensen, Ill. (Rag et al., 1976)	100	100	100	0	100	100	0	0	0	0	0
Yuma, Ariz.-Wellington, Calif. (Barmore, 1980)	100	100	100	100	100	100	0	0	0	0	0
Pecos, Tex. (Rives, 1980)	100	100	100	100	100	100	0	0	0	0	0
Russell, Kans. (Jantz et al., 1982)	100	100	100	100	100	100	0	0	0	0	100
Catron, N.M. (Johnson, 1985)	100	100	100	100	100	100	0	0	0	0	100
Cedar, Nebr. (Milliron, 1985)	100	100	100	100	100	100	0	0	0	0	100
Angeles, Calif. (Ryan and Giger,1988)	100	100	16	0	15	0	16	6	16	8	14

NOTE: AWC, Available water capacity; Exc. Base, exchangeable bases; CEC, cation-exchange capacity; Base Sat., base saturation.

TABLE 4-4 Land Areas Covered by Soil Surveys in 13 Western States (thousands of acres)

State	Total Surface Area[a]	Total Area of Rangelands[b]	Total of All Lands Mapped[c]	Percentage of Total Surface Area Mapped
Arizona	72,960	45,168	46,210	63
California	101,571	43,039	72,017	71
Colorado	66,618	27,821	60,766	91
Idaho	53,481	23,598	34,487	64
Montana	94,109	53,334	72,161	77
Nevada	70,758	56,887	55,417	77
New Mexico	77,819	48,725	68,634	88
North Dakota	45,249	12,295	42,970	95
Oregon	62,126	22,322	34,169	55
South Dakota	49,354	23,397	49,335	100
Utah	54,335	29,701	46,058	85
Washington	43,608	7,895	35,779	82
Wyoming	62,598	46,896	39,371	63

[a]Total surface area includes lands managed by USFS, BLM, and other federal agencies as well as nonfederal lands.

[b]Total surface area does not include areas of open water (U.S. Department of Agriculture, U.S. Forest Service, 1980).

[c]Soil Conservation Service compilation 1991. SCS Office, P.O. Box 2890, Washington, DC 20013. Acres reported include those covered by published surveys as well as surveys in progress.

soil are used and reused by plants, animals, and microorganisms. The amount of nutrients available and the speed with which nutrients cycle between plants and the soil are ecological processes fundamental to rangelands. Similarly, the total amount and time of year during which photosynthesis occurs are important indicators of how well rangeland ecosystems are functioning.

Nutrient Cycling

Nutrients follow cyclical patterns as they are used and reused by living organisms. Although the majority of available nutrients for plant growth is found in the soil (Brady, 1990), carbon, oxygen, and some of the nitrogen needed by plants are extracted from the atmosphere.

Nutrient cycling is closely related to the soil-water relationship on rangelands. Wind and water erosion strips away the nutrients stored in the topsoil (Logan, 1990). On most rangelands in the western United States, water is the most limiting factor for plant production. Generally, the greater the amount of precipitation that falls and that is captured and stored for later use by plants, the greater the total production of plant biomass on a particular rangeland site (Rauzi and Fly, 1968). Sites with high levels of biomass production more effectively capture and cycle the available nutrients compared with sites with lower levels of biomass production.

The total quantity of biomass produced, and hence, the total quantity of nutrients being cycled, also depends on the duration of the growing season. The longer the portion of the year in which plants are growing, the greater the total amount of biomass that can be produced. The length of the effective growing season is primarily determined by the amount and distribution of precipitation and temperature, but it can also be influenced by the particular composition of plants on a site. A plant that is photosynthetically active throughout the growing season, or a mixture of plants with various seasonal growth patterns, for example, may more effectively cycle the available nutrients as compared to a plant that is or a mixture of plants that are photosynthetically active for only a part of the growing season. The presence of actively growing plants during the entire growing season may indicate more complete capture and utilization of available nutrients.

Similarly, the degree to which the available root zone is occupied by plant roots may suggest the degree to which nutrients are utilized and cycled. Nutrient cycling entails the extraction of nutrients from the soil by plant roots. Rangelands may support individual plants with root systems that occupy much of the soil profile or a mixture of different plants that have various root depths (Figure 4-3), thus resulting in more complete utilization of the water and nutrients available throughout the entire soil profile (Table 4-5) (Weaver, 1954).

The organic materials, such as plant litter or animal feces, deposited on the soil surface are decomposed and reincorporated into the soil, and through nutrient cycling processes, they again become available to plants and other organisms. This decomposition and reincorporation of organic materials can be accomplished in a number of ways. Microorganisms (such as fungi and bacteria) and microinvertebrates (such as arthropods and nematodes) facilitate most of the decomposition of organic material as well as the formation of soil organic matter from partially decomposed organic residues (Paul and Clark, 1989).

Physical processes such as the action of wind, water, and sunlight are nonbiological means whereby plant parts are decomposed. These processes, combined with cycles of freezing and thawing or wetting and drying, leach or physically remove the nutrients in plant materials (Laycock and Price, 1970). Fire can rapidly release the nutrients immobilized in plant tissues. These nutrients may be carried away in smoke and ash or

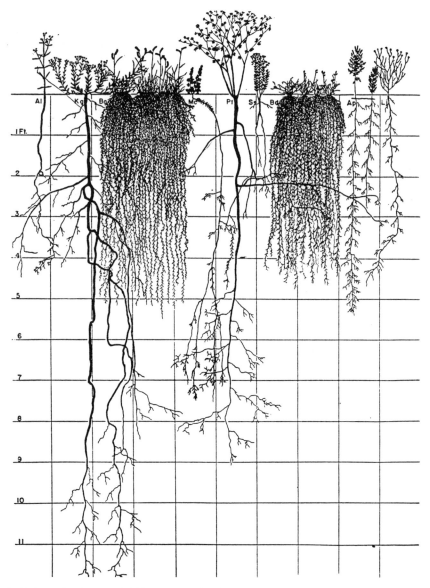

FIGURE 4-3   Roots of different grassland plants draw their moisture from different soil layers. Roots of some native plants extend to depths of 20 or more feet. Al, narrow-leafed 4-o'clock (*Allionia linearis*); Kg, prairie false boneset (*Kuhnia gultinosa*); Bg, blue grama (*Bouteloua gracilis*); Mc, globemallow (*Malvastrum coccineum*); Pt, a legume (*Psoralea tenuiflora*); Ss, *Sideranthus spinulosis*; Bd, buffalo grass (*Buchloe dactyloides*); Ap, western ragweed (*Ambrosia psilostachya*); and Li, skeleton weed (*Lygodesmia juncea*). Source: A. Stefferud, ed. 1948. Grass: The Yearbook of Agriculture 1948. Washington, D.C.: U.S. Department of Agriculture.

TABLE 4-5 Rooting Depth in Prairie Soils

Layer	Depth (m)	Percentage of Species	Examples
Shallow	0.6	14	Blue grama (*Bouteloua gracilis*), Prairie June grass (*Koelaria cristata*)
Medium	0.6–1.5	21	Needle grass (*Stipa spartea*), Buffalo grass (*Buchloe dactyloides*), Many-flowered aster (*Aster ericoides*)
Deep	1.5–6.0	65	Big bluestem (*Andropogon gerardi*), Slough grass (*Spartina pectinata*), Compass plant (*Silphium laciniatum*)

SOURCE: R. Brewer. 1989. The Science of Ecology. Philadelphia: Saunders College Publishing. Data are based on J. E. Weaver and F. E. Clements. 1938. Plant Ecology, 2nd ed. New York: McGraw-Hill, and other publications by J. E. Weaver.

may be deposited on the soil surface as readily available nutrients (Christensen et al., 1989).

The rate at which nutrients are cycled and the total volume of nutrients in a state of transition (flux) are important processes of rangeland ecosystems. The capacity of rangelands to produce resources and satisfy values depends on the buildup and storage of nutrients over time. Interruption or slowing of nutrient cycling can lead to site degradation as a rangeland becomes increasingly deficient in the nutrients required by plants. An evaluation of the degree to which nutrients are conserved and the degree to which nutrient cycles operate should be important elements of a system of assessing rangeland health.

Energy Flow

Nearly all living things depend on the process of photosynthesis, by which green plants capture energy and convert it to chemical energy, which is then stored in the plant. The amount and timing of the sunlight that reaches any point on the ground are influenced by a number of factors, including latitude, elevation, and weather. Green plants trap and process the solar energy. The capture of energy, however, requires the expenditure of energy, which is in turn lost to the ecosystem as heat. Thus, energy flows rather than cycles through the ecosystem, and the ecosystem is dependent on green plants to continually capture the sun's energy.

On rangelands, grasses, forbs (an herb other than a grass), shrubs, and trees are the primary converters of sunlight energy. The energy is converted by these plants and is stored primarily as carbohydrates. This energy supports most other life-forms. Herbivores harvest the plant material, and through the process of digestion, they gain the energy stored in the plant tissues (Holechek et al., 1989). The energy that is captured and converted by green plants moves through the ecosystem in a process that can be described as an energy pyramid (Figure 4-4). The base is the energy from photosynthesis stored in plants. At each level, organisms use some energy for maintenance, and this energy is given off in respiration and heat, resulting in a highly inefficient process (Brewer, 1989).

The total volume of sunlight energy that is captured is an important determinant of the resources and values produced by rangelands. Rates of energy flow can vary on rangelands in both space and time. Indicators of change in the spatial and temporal distributions of energy flows should be an important component of a comprehensive system of evaluating rangeland health.

DISTRIBUTION OF ENERGY FLOW OVER TIME

Energy flow can be affected in a number of ways. Since energy can be converted by plants only when they are actively green and growing, anything that affects this amount of time will affect the energy flow. Plant life-forms and species compositions determine the ability of the plant community to process sunlight energy under a variety of environmental conditions. Plant interactions with the physical environment also influence energy flow. For example, as plant cover increases, the effectiveness

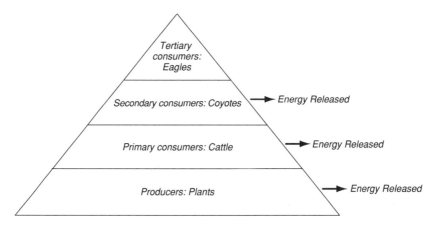

FIGURE 4-4 The loss of energy as it moves through trophic levels. Source: Derived from B. J. Nebel. 1981. P. 33 in Environmental Science: The Way the World Works. Englewood Cliffs, N.J.: Prentice-Hall. Reprinted, with permission, from Bernard J. Nebel (1981). © 1981 by Prentice-Hall, Englewood Cliffs, N.J.

of rangelands in capturing and storing precipitation often increases. Increased plant cover, however, may reduce soil temperatures, thereby reducing the effective length of the growing season.

Many different physiological adaptations and life history strategies of different plant species influence the amount of energy that is captured. A plant that possesses a C_4 photosynthetic pathway (warm season) uses water more efficiently than does a plant with a C_3 photosynthetic pathway (cool season) (Fitter and Hay, 1987). Rangelands with both warm- and cool-season plants can effectively extend the period during which photosynthetic activity occurs.

SPATIAL DISTRIBUTION OF ENERGY FLOW

The energy flow within an ecosystem also varies spatially at the level of an individual plant and at the level of the community. Individual plants have different growth forms (the shape and arrangement of leaves, stems, and branches), and solar energy capture is influenced by these growth forms. The vertical structure of plant growth creates layers in the plant canopy, with plants adapted to the different quantities of light available within each layer.

Criteria and Indicators of Nutrient Cycling and Energy Flows

Nutrient cycling and energy flow have been studied as part of efforts to understand the functioning of rangelands. An evaluation of nutrient cycles and energy flows, however, has not been part of traditional assessments of rangelands. Therefore, experience with useful criteria and indicators of nutrient cycles and energy flow on rangelands is much less than that with indicators of soil stability and watershed function. Current knowledge can be used as a starting point for the development of useful criteria and indicators of nutrient cycling and energy flow.

The distributions of nutrients and energy in space and time should be the criteria used to evaluate the integrity of nutrient cycles and energy flow.

Plants depend on the nutrients in the soil and energy captured from the sun. Nutrients stored in the soil are used and reused by plants, animals, and microorganisms. The amount of nutrients available and the speed with which nutrients cycle between plants, animals, and the soil are fundamental processes of rangelands. Similarly, the amount, timing, and distribution of energy captured through photosynthesis are fundamental to the function of rangeland ecosystems. Indicators that can be used to evaluate the spatial and temporal distributions of nutrients and energy should be part of a comprehensive evaluation of rangeland health.

COMMUNITY STRUCTURE

The structure of plant communities—that is, the growth habits, life-forms, and distribution of species—may be a useful indicator of nutrient cycling and energy flow. The suggested indicators summarized in Table 4-2 emphasize the importance of a community structure in which the available niches are filled on both a spatial and a temporal basis. The proportion of the growing season during which plants are photosynthesizing may prove a useful indicator of the of the amount of available energy captured by rangeland ecosystems. Since the growing season is determined by both temperature and rainfall, the total growing season in arid rangelands may be the sum of periods following episodic rainfall when moisture is available for plant growth. Data on species composition could be used to assess whether photosynthesis occurs during all or only a part of the growing season. Similarly, data on plant composition could used to assess whether all or only a fraction of the available soil profile is occupied by plant roots. Change in a community structure that results in shortening of the period during which photosynthesis occurs or in utilization of a smaller volume of the soil profile for extraction of nutri-

Researchers study root production plots labeled with ^{14}C, providing important new insights into below-ground processes. Credit: Long-Term Ecological Research Network, University of Washington.

ents and water could be interpreted as indicating a decrease in the total amount of nutrients cycled and in energy flow through the community. Most undegraded rangelands include a mix of plants that, in total, are capable of effectively utilizing nutrients and carrying out photosynthesis during the growing season. Examples of rangelands with such a mix of plants are the sagebrush-grass ecosystems in Wyoming and the prairies of the Great Plains. There are, however, rangelands such as the blue grama grasslands of the western Great Plains and alkali sacaton grasslands in moist, saline sites, where a single species dominates the plant community, yet the available nutrients and growing season are still effectively utilized. In other cases, the domination of the plant community by a single species indicates that some niches are not filled and the available nutrients and growing season are not being effectively utilized. Rangelands dominated by cheatgrass, a shallow-rooted plant that grows for only a short period of time in the spring and fall, are one example. Sagebrush-grass ecosystems that have lost most of the grass component because of overgrazing are another example.

DISTRIBUTION OF LITTER AND PLANTS

The degree of fragmentation of nutrient cycles and energy flows as indicated by the pattern in which litter and plants are distributed across the site may also serve as a useful indicator of rangeland health. A fragmented distribution of litter and plants in which there are large bare areas interspersed between patches of litter and rooted plants seems to indicate unfilled niches and provides opportunities for erosion to occur. Unfilled niches suggest that opportunities for plants to capture sunlight through photosynthesis and contribute to the biomass production of the site have been lost. Erosion from bare areas represents a loss of nutrients from part of the site. Such nutrients may move only a short distance before they are trapped by vegetation and other obstacles, but such movement may represent the increasing vulnerability of a site's nutrient cycles to interruption.

The secretaries of the U.S. Department of Agriculture (USDA) and of the U.S. Department of the Interior (DOI) should initiate a coordinated research effort, drawing on federal agency and other scientists to develop, test, and implement indicators of the spatial and temporal distributions of nutrients and energy for use in rangeland health assessments.

The lack of experience with and testing of specific indicators of nutrient cycling and energy flow is an important impediment to the development of a comprehensive system of determining whether rangelands are healthy, at risk, or unhealthy. There is an urgent need for basic and applied research to develop useful indicators and the understanding needed to interpret the significance of changes in those indicators. The development of criteria and indicators that can be used to evaluate nutrient cycling and energy flow in rangeland ecosystems is essential for evaluating rangeland health and for proper rangeland management.

JUDGMENT REQUIRED

Until measurable indicators of nutrient cycling and energy flow are developed, evaluation of what constitutes a healthy, at-risk, or unhealthy distribution of plants, bare areas, rooting depths, and growth periods will depend primarily on informed judgments. The healthy end of the continuum consists of an unfragmented distribution of plants and litter with few bare areas, plants that fill the soil profile with roots, and plants that are capable of photosynthesis throughout the growing season. The unhealthy end of the continuum probably consists of a fragmented plant cover with many large bare areas, plants that fill only a small portion of the soil profile with roots, and plants that are capable of photosynthesis during only a short portion of the growing season.

RECOVERY MECHANISMS

For a rangeland to maintain a healthy state or naturally evolve toward a more healthy state, mechanisms that allow such an evolution on the site must be in place and they must be working. Recovery mechanisms generally involve extension of biotic control over the abiotic environment through the processes of soil and plant community development.

Functioning recovery mechanisms that lead to capture and cycling of nutrients, capture of energy, conservation of nutrients, energy, and water within the site and to development of resistance and resilience to extreme events such as drought, fire, or rainstorms are fundamental to rangeland ecosystems. Indicators of change in the operation of recovery mechanisms should be an important component of a comprehensive system of evaluating rangeland health.

Criteria and Indicators of Recovery Mechanisms

Changes in plant demographics should be the criterion used to evaluate recovery mechanism activities.

Experience with and testing of indicators of the functioning of recovery mechanisms are limited. Various indicators of plant demographics, however, have been commonly used as measures of apparent trend (SCS). Indicators of age class distribution, plant vigor, and the presence and distribution of microsites for seed germination and seedling development would be useful starting points for the development of more systematic indicators of the function of recovery mechanisms on rangelands.

AGE-CLASS DISTRIBUTION

A measure of the age-class distribution of the plant species present on a rangeland site has been used in past rangeland evaluations. Such evaluations attempt to show whether the plant species are replacing themselves and whether a plant community is expected to maintain control of the site. Interpretation of age-class data can be difficult in some situations. In arid lands, for example, seedling establishment is often episodic—a large number of seedlings are established following episodic rainfall events, while few seedlings are established during the long periods when no rainfall occurs. Lack of young plants may reflect the vagaries of climate more than the health of the rangeland. In many cases, however, the lack of plants of certain age classes or a predominance of old or deteriorating plants may indicate a change in the plant community's structure and function. Range managers have previously used this concept to judge apparent trend (SCS). However, an upward trend in range condition (SCS) was almost always based on whether seedlings of plants that are part of the climax plant community (SCS) were present. In assessing rangeland health, when judgments are used to place a higher value on one plant than another, emphasis should be placed on species that reduce soil erosion or fill nutrient cycling or energy flow niches. Such decisions must be made on a site-by-site basis.

PLANT VIGOR

The vigor of the vegetation present has also been used in the past to assess the dynamics of plant communities. Indices that have been used to judge plant vigor are color, seed and rhizome production, size of plants, and annual amount of biomass produced. To the extent that vigor can be detected and defined, it is a useful tool because a decline in plant vigor precedes changes in plant species composition, the development of bare areas that are susceptible to soil erosion, and the development of open niches. Determination of vigor is largely subjective, however, since there are no precise determinants. The following indicators, for example, were suggested by Stoddart and Smith (1955) to assess the effect of grazing on plant vigor:

Class 1. Palatable plants are vigorous. Grasses are robust with numerous leaves, seedstalks are tall and numerous, and leaves are dark green. There are no hedged or high-lined browse (tender shoots, twigs, and leaves of trees and shrubs used by animals for food). Forage plants reproduce. Rating = 10 points.

Class 2. Palatable plants lack vigor. Forage grasses are shorter, and there are fewer seedstalks than in class 1. Seedlings may be present. Few forage plants of younger age classes are represented. Less-palatable weeds and grasses are generally vigorous. Rating = 7 points.

Class 3. Palatable plants lack vigor. Grasses are weak. Forage plants do not reproduce. Rating = 5 points.

Class 4. Palatable plants are sickly and weak. Grasses may be pale or yellowish, seedstalks are few and short, and there are no seedlings. Palatable plants do not reproduce, and sod is thinning. Rating = 1 point.

Evaluation of plant vigor may be a useful indicator of changes in the function of recovery mechanisms on rangelands. It is important, however, that an evaluation of rangeland health not be confused with determination of whether a rangeland is valuable for a particular use. The rangeland health criteria suggested for plant vigor, therefore, should not emphasize palatable forage plants or reflect implicit judgments on the relative value of grass, browse, or weed species. Evaluation of plant vigor requires knowledge of plants, rangeland ecology, and site characteristics. Characteristics such as a mix of plants with normal growth on the basis of height, color, seed production, rhizome and stolon production (rhizomes and stolons are modified stems that help a plant spread laterally from a parent plant), and annual biomass production may prove useful indicators.

MICROSITES FOR SEED GERMINATION AND
SEEDLING ESTABLISHMENT

Finally, maintenance of biotic control over the abiotic environment (the nonliving part of the environment) through self-induced changes in plant community dynamics requires the presence of microsites (the area immediately surrounding a seed, which may be as small as a few millimeters in diameter) that are favorable for seed germination and seedling establishment because of increased moisture, nutrients, and protection from herbivory (Harper, 1977). Competition from existing plants, soil erosion by wind or water, and the development of soil crusts are important processes that affect germination and seedling establishment on microsites. Indicators of the availability of microsites should be developed to serve as useful indicators of changes in the function of recovery mechanisms.

RESEARCH NEEDED

The secretaries of USDA and DOI should initiate a coordinated research effort, drawing on federal agency and other scientists to develop, test, and implement indicators of recovery mechanisms for use in rangeland health assessments.

Explicit evaluations of recovery mechanisms have not been part of rangeland assessments to date. The lack of development and testing of measurable indicators of change in recovery mechanisms is an important impediment to a comprehensive evaluation of rangeland health. There is an urgent need to develop measurable indicators and methods of evaluation that can be incorporated into routine assessments of rangelands. The indicators discussed here are embryonic and are suggested by the committee to stimulate research, development, and refinement of useful indicators of self-induced recovery mechanisms.

MEASUREMENT AND EVALUATION OF INDICATORS OF RANGELAND HEALTH

The indicators of rangeland health for each of the criteria listed in Table 4-2 are only a subset of those that the committee could suggest. Past investigators have used all of the indicators as part of rangeland assessments. In many cases, the data currently collected as part of range condition (SCS), ecological status (USFS and BLM), and apparent trend (SCS) evaluations can be reinterpreted as indicators of rangeland health. The overlap between the indicators of rangeland health investigated by the committee and currently used indicators is given in Table 4-6.

Identification of Boundaries

A fundamental problem with assessing rangeland health involves identification of the boundaries between healthy, at-risk, and unhealthy rangelands. Table 4-7 illustrates one way these boundaries could be identified using indicators of soil stability and watershed function, distribution of nutrients and energy, and recovery mechanisms—the three-phase approach recommended by the committee. The illustration in Table 4-7 and the discussions that follow are offered as a useful starting point. These distinctions will have to be refined and validated through research.

SOIL STABILITY AND WATERSHED FUNCTION

Healthy rangelands, in this example, exhibit no evidence of accelerated soil erosion by wind or water. There is no evidence of the formation of rills and gullies, pedestals, or sheet or scour erosion, and there is no evi-

TABLE 4-6 Indicators of Rangeland Health Currently Used as
Indicators of Range Condition (SCS), Ecological Status (USFS and
BLM), or Apparent Trend (SCS)

	Same or Similar Indicator Used for	
Proposed Rangeland Health Indicator	Range Condition (SCS) and Ecological Status (USFS and BLM)	Apparent Trend (SCS)
A-horizon		
Rills, gullies		DM
Pedestaling		DM
Scouring or sheet erosion		DM
Sedimentation, dune formation		RM
Plant distribution		
Litter distribution and incorporation		RM
Rooting depth	RM	
Photosynthetic period	RM	
Age class		RM
Plant vigor		DM
Germination microsites		

NOTE: DM, direct measure of indicator used; RM, related measure of indicator used.

dence that water- or wind-eroded materials have been deposited on the
site. The A-horizon of the soil appears to be stable and is present uni-
formly over the site.

At-risk rangelands show evidence of soil movement, but such movement is primarily within the site itself. Rills and gullies may be forming,
but they are not yet well developed or integrated into a dendritic pattern.
If pedestaling is present, it is not so severe that the roots are exposed.
Similarly, scours and dunes are small and not well developed, if they are
present, and there is little evidence of sediment deposits on the site. Soil
particles, organic matter, nutrients, and water are redistributed on the
site, but they are not yet lost from the site.

Evidence of soil movement off the site, however, indicates an unhealthy state. Rills and gullies are well developed and active, and they
display a developed dendritic pattern. Pedestaling is severe enough that
roots are exposed, scours and dunes may be active and widespread, and
large areas may be devoid of the A-horizon of soil. There is clear evidence of soil degradation and transport of nutrients, water, and organic
matter off the site.

DISTRIBUTION OF NUTRIENTS AND ENERGY

In Table 4-7, the nutrients stored in litter, plant, and root biomass are well distributed and litter is being decomposed and incorporated into the soil throughout the rangeland considered to be healthy. Photosynthetic activity, which is represented by actively growing plants, is also well distributed across the site. The rangeland considered at risk, however, shows evidence that the spatial and temporal distribution of nutrients and energy is becoming fragmented across the site. Litter may be present, but it tends to accumulate in depressions or around prominent grasses or shrubs. The plant and root biomass is beginning to show a fragmented pattern, and barren areas develop between patches. In the rangeland considered unhealthy, this fragmented distribution of nutrients and energy is pronounced. Litter is sparse, accumulating and being incorporated

TABLE 4-7 Relationship between Health Criteria and Thresholds

Phase	Healthy	At Risk	Unhealthy
1. Soil stability and watershed function	No evidence of soil movement	Soil is moving, but remains on site	Soil is moving off site
2. Distribution of nutrients and energy	Plant and litter distribution unfragmented	Fragmented distribution developing	Fragmented distribution developed, with large barren areas between fragments
	Photosynthetic activity occurs throughout the period suitable for plant growth	Photosynthetic activity restricted during one or more seasons	Photosynthetic activity restricted to one season only
	Rooting throughout the available soil profile	Roots absent from portions of the available soil profile	Rooting in only one portion of the available soil profile
3. Recovery mechanisms	Diverse age-class distribution Plants are vigorous	Seedlings and young plants are missing Plant vigor is reduced	Decadent plants predominate Plant vigor is poor
	Germination microsites are present and well distributed	Developing crusts or soil movement degrade microsites	Soil movement or crusting inhibit most germination

into the soil only in depressions or around prominent grasses or shrubs. Plant and root biomass is restricted to patches, and there are large barren areas between the patches.

The distribution of nutrients and energy over time is also used to distinguish between states in Table 4-7. Photosynthetic activity occurs during most of the period when temperature and moisture make photosynthesis possible in rangelands defined as healthy. The diversity of plants with different growing periods is sufficient to ensure that at least some plants are actively photosynthesizing during most of the period when the temperature and moisture are at levels such that photosynthesis can occur. Similarly, the plant community structure supports a diversity of rooting depths, so that plants use the nutrients and water available throughout the soil profile. In the at-risk category, the period of time when photosynthesis occurs is reduced, and the portion of the soil profile occupied by the roots of the plants that are present is restricted. In the unhealthy category, photosynthesis is restricted to a portion of the period when moisture is available, and the plant roots occupy only one layer in the soil profile.

RECOVERY MECHANISMS

In Table 4-7, rangelands that show evidence that plant community dynamics are sufficient to at least maintain the current community structure and function are classified as healthy. There is a diverse species composition and age-class distribution, microsites in which seeds can germinate are available, and seedlings are becoming established. Plants are vigorous and show no signs of deformed growth patterns. At-risk rangelands are distinguished by missing age classes, reduced plant vigor, and restricted seedling recruitment. Rangelands in the unhealthy category are characterized by the predominance of old or deteriorating plants, the loss of microsites for seed germination, and the absence of seedlings.

Preponderance of Evidence from Measured Indicators

The decision to classify a rangeland as healthy, at risk, or unhealthy should be a judgment based on the preponderance of evidence from an evaluation of multiple and measurable indicators. Data must be collected and reported for all the measurable indicators to allow for independent analyses of the final classification decision. Consistently reported data on measurable indicators will be more valuable than a final judgment.

It is unreasonable to expect that all indicators will simultaneously fall into the healthy, at-risk, or unhealthy category. It is more likely, for example, that the soil of a particular rangeland may show no evidence of move-

ment by wind and water, suggesting that the site is healthy, while the distribution of litter, plants, and roots across the site may be becoming patchy, suggesting that the site is moving toward the at-risk category. In such cases a judgment must be made. The development of a patchy distribution of litter and plants suggests a reduction in the effective soil cover, which may lead to accelerated erosion or increased vulnerability to serious soil erosion during an intense rainstorm, and so the site is best considered at risk. Alternatively, seed germination and seedling recruitment may indicate that the patchy distribution will be short-lived, and so the site is best considered healthy. Each indicator represents an important piece of the puzzle. Taken together they provide the information needed to assess rangeland health.

New Models of Rangeland Change Needed

The secretaries of USDA and DOI should initiate a coordinated research effort, drawing on federal agency and other scientists to develop, test, and implement new models of rangeland change that incorporate the potential for difficult-to-reverse shifts across ecological thresholds.

A theory is accepted until unexplained anomalies overwhelm the proponents of the current theory or until an alternative theory is proposed to explain the anomalies. Recent findings in the field of community ecology have questioned the applicability of climax theory to all rangelands. Laycock (1989), Westoby (1980), Westoby and colleagues (1989), and others have reviewed the anomalies that are not well explained by successional theory. It still appears, however, that the area in the United States containing anomalous grassland vegetation is smaller than the area where successional theory can be used to model plant community dynamics.

New models are being proposed to better explain the dynamics of rangeland vegetation. Researchers in arid areas, notably, in Australia and South Africa, are advancing new theories to explain the responses of plants to grazing and other environmental factors. Concepts of population dynamics that are derived from the field of animal ecology are being applied to rangeland vegetation, and concepts of alternating stable communities are challenging the Clementsian model of successional change (Clements, 1916). A state and transition model proposed by Westoby and colleagues (1989) seems well adapted to rangelands where episodic events may well be the primary factors responsible for determining vegetation composition.

Despite these advances, there does not appear to be a single coherent theory that can explain all of the current anomalies or that has been sufficiently tested to replace current successional concepts. The facts that current successional theory apparently adequately explains vegetation dynamics on a significant portion of rangelands and that no new comprehensive theory has yet emerged have restrained efforts to replace the successional models currently used to rate range condition (SCS) or ecological status (USFS and BLM).

The development of new models and theories of rangeland change requires research. There is a need to fund such research at an interdisciplinary level, integrating range science theories with theories from the other ecologies. The drive to specialize is not unique to range science or ecological research; however, this does impede the transfer of new and possibly helpful ideas between specialized fields. In addition, current ecological research is oriented toward the development of new ideas rather than the testing of new ideas for a variety of rangeland types. This testing of new ideas is essential to determine which new ideas might hold promise as a theoretical foundation for rangeland management. The lack of a single, coherent theory of community structure and development that can replace current climax, succession, and retrogression models is the result of the fact that such potential theories cannot be sufficiently tested on a sufficiently large number of sites and, in so doing, allow researchers to gain confidence that such theories could replace the currently held ones.

There is a need for inexpensive inventory, classification, and monitoring methods with links to current ecological theory. These links must be robust so that as ecological theories change, the data can still be interpreted using new theories. This will involve a multiple-attribute approach to the design of the inventory.

The development of such methods will require funding of interdisciplinary research that links all branches of ecological research. Furthermore, research to demonstrate the applicability of newly developed models over a variety of rangeland types is required. This area of research must be the testing ground for models that might hold promise as an improved theoretical framework for classifying, inventorying, and monitoring rangelands.

Better Understanding of Rangeland Soils Needed

The secretaries of USDA and DOI should initiate a coordinated research effort, drawing on federal agency and other scientists to increase understanding of the relationship between soil properties and rangeland health.

Knowledge of soils has been used in assessments of rangelands primarily as an aid in the classification of rangelands. A range site (SCS), ecological site (BLM), or ecological type (USFS) is often defined by the presence of one or more characteristic soil types.

While much research and experience supports the relationship of soil surface characteristics to rangeland health, basic knowledge of the effects of other soil properties such as organic matter content or water-holding capacity on nutrient cycling, energy flow, recovery mechanisms, and other elements of rangeland health is limited. The effects of grazing management and other management practices on soil properties are also not well understood. Basic and applied research is needed to increase understanding of how changes in soil properties affect rangeland health.

Research on cropland soils has attempted to identify those soil properties that are most important in determining crop yield and vulnerability to degradation. Much research has also been devoted to determining the impact of tillage, crop sequence, residue management, organic amendments, fertilizers, cover crops, and other elements of farming systems on soil properties.

Research pertaining to the coupling of rangeland soils with rangeland vegetation and disturbance of the rangeland ecosystem is needed. Although recent work has addressed the influence of soil organic matter on rangeland productivity (Burke et al., 1989), additional research detailing soil organic matter at specific sites is necessary to more closely describe the nutrient dynamics of range sites.

Study of the relationships of other soil properties to plant growth and species composition is limited in the literature on rangeland soils. Some reports have addressed this area (Passey et al., 1982), but many questions remain. For example, at what level do plant types (species, community, landscape) and soil types (series, family, subgroup) most closely correlate?

The influence of soil degradation on plant parameters (for example, yield, species composition, and regenerative power) is not well researched for rangeland soils. The influences of plowing or soil removal, grazing intensity, and climatic degradative factors (water and wind erosion) have not been determined for rangelands. All of these questions have a direct bearing on rangeland health.

Field Evaluation of Proposed Indicators

To explore the usefulness of the criteria and indicators discussed above, the committee developed a three-phase matrix as a workable approach to implementing its recommended three-phase concept of rangeland health evaluation. The matrix is presented in Table 4-8. The committee attempted to develop measures of the indicators that involve simple and, in many cases, visual estimates. The committee conducted limited field tests of a three-phase evaluation of rangeland health using the matrix.

Eighteen limited field tests were carried out by six members of the

TABLE 4-8 Rangeland Health Evaluation Matrix

Indicator	Healthy	At Risk	Unhealthy
		Phase 1: Soil stability and watershed function	
A-horizon	Present and distribution unfragmented	Present but fragmented distribution developing	Absent, or present only in association prominent plants or with other obstructions
Pedestaling	No pedestaling of plants or rocks	Pedestals present, but on mature plants only; no roots exposed	Most plants and rocks pedestaled; roots exposed
Rills and gullies	Absent, or with blunted and muted features	Small, embryonic, and not connected into a dendritic pattern	Well defined, actively expanding, dendritic pattern established
Scouring or sheet erosion	No visible scouring or sheet erosion	Patches of bare soil or scours developing	Bare areas and scours well developed and contiguous
Sedimentation or dunes	No visible soil deposition	Soil accumulating around plants or small obstructions	Soil accumulating in large barren deposits or dunes or behind large obstructions
		Phase 2: Distribution of nutrient cycling and energy flow	
Distribution of plants	Plants well distributed across site	Plant distribution becoming fragmented	Plants clumped, often in association with prominent individuals; large bare areas between clumps

		Phase 3: Recovery mechanisms	
Litter distribution and incorporation	Uniform across site	Becoming associated with prominent plants or other obstructions	Litter largely absent
Root distribution	Community structure results in rooting throughout the available soil profile	Community structure results in rooting in absence of roots from portions of the available soil profile	Community structure results in rooting in only one portion of the available soil profile
Distribution of photosynthesis	Photosynthetic activity occurs throughout the period suitable for plant growth	Most photosynthetic activity occurs during one portion of the period suitable for plant growth	Little or no photosynthetic activity on location during most of the period suitable for plant growth
Age-class distribution	Distribution reflects all species	Seedlings and young plants missing	Primarily old or deteriorating plants present
Plant vigor	Plants display normal growth form	Plants developing abnormal growth form	Most plants in abnormal growth form
Germination microsite	Microsites present and distributed across the site	Developing crusts, soil movement, or other factors degrading microsites; developing crusts are fragile	Soil movement or crusting sufficient to inhibit most germination and seedling establishment

committee in Arizona, New Mexico, Colorado, Wyoming, Montana, and Nebraska between April and July 1991. An analysis of the results revealed three important points in evaluating rangeland health by the approach proposed by the committee.

SOIL STABILITY AND WATERSHED FUNCTION ARE THE MOST IMPORTANT CRITERIA

Soil stability and watershed function should have greater weight than other criteria in the determination of rangeland health; soil movement off site should mean the rangeland is unhealthy.

The degradation of soil and watershed function that accompanies movement of soil by wind and water can lead to irreversible changes in productivity and site potential, at least within a practical time scale and within the realm of economically feasible reclamation efforts. The irreversibility of such erosion losses has been repeatedly cited as a major concern in croplands, and forests as well as rangelands (Bormann and Likens, 1979; Ellison, 1949; Klock, 1982; Larson et al., 1983; Pierce, 1991; Sheridan, 1981; Wight and Siddoway, 1982).

Because of the danger of irreversible effects from soil and watershed degradation, the evaluation of soil stability and watershed function should have greater weight in the final evaluation of rangeland health than do the other two phases of evaluation. The result of the phase 1 evaluation should be the maximum rating a rangeland can receive. That is, if an evaluation of phase 1 results in an unhealthy rating, evaluation of the other components should not result in a final rating of at risk or healthy.

Uncertainty in Interpreting Indicators

The precise placement of the boundaries between healthy, at risk, and unhealthy, however, is not clear. Placement of rangelands into healthy, at-risk, or unhealthy classes will require judgment. As in any classification scheme, there will be borderline cases that are difficult to place into one class or another.

The scientific understanding required to develop and interpret changes in indicators of rangeland health is currently better developed for phase 1 than it is for phases 2 and 3. There is substantial experience in using soil surface characteristics as indicators of soil stability, and new models of soil erosion on rangelands are being developed. The addition of currently used evaluations of soil surface characteristics, as described in this report, to all current and ongoing range condition (SCS), ecological status (USFS and BLM), and apparent trend (SCS) assessments would be a useful first step toward an assessment of rangeland health—a step that could be taken immediately.

The scientific understanding needed to develop and interpret measurable indicators of changes in nutrient cycling, energy flow, and self-induced recovery mechanisms is less well developed—an effort to develop and test such is urgently required.

Finally, the difficulty in standardizing the boundaries between states of rangeland health must not impede efforts to expand the data collected during rangeland inventorying and monitoring. Standardizing rangeland health assessments will require systematically assembled data on measurable indicators. Collection of such data will allow tests of the utility of such measures in indentifying healthy, at-risk, and unhealthy rangelands.

5 Inventorying and Monitoring Rangeland Health

Rangeland inventories have been initiated for a variety of reasons: scientific curiosity, economic necessity, and legal mandates. Ownership patterns in the western United States have influenced the demand for and the type of information needed from rangelands. Increased awareness of environmental issues has also influenced the type of information demanded for range management. These different incentives have resulted in a variety of inventory methods at a variety of spatial scales. Site-specific inventories, subject-specific inventories, and national inventories currently exist.

PAST INVENTORIES OF RANGELANDS

The first recorded surveys of U.S. rangelands are fragmentary comments from the letters and journals of explorers, trappers, members of military expeditions, and missionaries. They are based on visual observations of what were, to them, vast and foreign landscapes. Early military and exploratory expeditions and the medical and scientific plant collectors who accompanied them provided much information about western U.S. rangelands (Chapline and Campbell, 1944). Although few of these people described the vegetation or the grazing resource sufficiently for a credible analysis of the soil and plant conditions prior to settlement by Europeans, their records served to entice additional surveys.

The First Surveys

Botanical surveys were formalized in the 1800s. Created in 1869, the Division of Botany of the U.S. Department of Agriculture (USDA) initiated a number of surveys of plants and catalogued the flora and poisonous plants of the western United States (Chapline and Campbell, 1944; Wass-

Sandberg bluegrass (*Poa secunda*)

134

er, 1977). During the same period, land surveyors were working on western lands, and some of their notes are fairly detailed in their descriptions of vegetation. The U.S. Bureau of the Census kept track of records of livestock populations across the United States.

During the 1800s, landholdings in the western United States were a changing mix of public domain and private land. Livestock owners developed operations based on the intermixes of these lands and the seasonal availability of forage. Western rangelands were being changed from vast and foreign landscapes to important commercial resources. Western livestock operations flourished during the period from 1867 to 1887 as railroads connected these grazing lands with the expanding cities of the Midwest and the East (Rowley, 1985).

In the late 1800s, the dramatic expansion of the livestock industry combined with equally dramatic droughts, and severe winters led to widespread degradation of rangelands. In 1895, letters from individuals about the declining range conditions in Texas were sent to USDA, and plans were made for an experiment station in Abilene, Texas. The Division of Agrostology within USDA was created in 1895 to investigate grass, forage, and range management (Wasser, 1977). Although it lasted for only 6 years, this division published a number of reports on the forage conditions and grazing problems of the western United States (Chapline and Campbell, 1944). These surveys were localized descriptions of rangeland vegetation and management.

FOREST RESERVE SURVEYS

In 1891, the U.S. Congress authorized the President to set aside national forests from the unreserved public domain.[1] This act was the first step in closing America's vast open land frontier. This act also marked a change in the kind of information the government needed about these lands—it shifted from landholding to land management information.

The management direction of these forest reserves was not officially set until 1897, when Congress decreed that national forests were established to "improve and protect the forest . . . or for the purpose of securing favorable conditions of water flows, and to furnish a continuous supply of timber."[2] In 1901, USDA developed principles to regulate grazing within U.S. forest reserves. Grazing was allowed "where it is shown after careful examination, that grazing is in no way injurious to or preventive of the conservation of the water supply" (Rowley, 1985:40).

In 1905, a letter from the secretary of USDA to the director of the newly created U.S. Forest Service (USFS) indicated that the rangelands were to be appraised as to their carrying capacity—that is, the maximum limit of livestock numbers that could be grazed on that land—to divide the land and set limits for grazing permits (Rowley, 1985). This need for information to determine grazing capacity led Jardine and others to develop methods of surveying forest reserves (Chapline and Campbell, 1944). These methods were, by necessity, limited because surveys of extensive areas were required. Although these determinations of carrying capacity were a regular responsibility of forest managers, only the annual numbers of permits and numbers of livestock that grazed the forest reserves were published in the annual reports of the USFS (Rowley, 1985).

STATE EXPERIMENT STATION SURVEYS

Around the early 1900s, states were establishing state experiment stations in response to the research and management needs of the agricultural industries within their borders. This activity also resulted in localized inventories of rangelands, such as the Washington Agricultural Experiment Station bulletin by John Cotton on the range conditions of central Washington State (Chapline and Campbell, 1944).

BLM GRAZING DISTRICT SURVEYS

In 1934, the U.S. Congress passed the Taylor Grazing Act,[3] thereby putting an end to the status of unreserved federal lands as unregulated common lands. The Taylor Grazing Act was intended to "stop injury to the public grazing lands by preventing overgrazing and soil deterioration, to provide for their orderly use, improvement, and development, to stabilize the livestock industry dependent upon the public range, and for other purposes."[4] Virtually all remaining land in the public domain— about 69 million hectares (170 million acres)—outside Alaska was withdrawn into grazing districts under the jurisdiction of the Bureau of Land Management (BLM) (Peffer, 1951; Voss, 1960). Although the act's preamble made it clear that rangeland improvement was a primary congressional goal, the U.S. Congress provided no precise guidelines in the act for achieving that aim. The Taylor Grazing Act addressed range inventories only indirectly by referring to grazing capacities.

SURVEYS OF NONFEDERAL RANGELANDS

In 1935, the U.S. Congress, responding to the devastation of the Dust Bowl of the early 1930s, established the Soil Conservation Service (SCS) as part of USDA to carry out programs for the control and prevention of soil erosion on public and private lands.[5] In enacting this legislation, which was known as SCS's "Organic Act," Congress officially recognized the need to gather soil, vegetation, and other resource data on nonfederal

Nearly 60 years ago, the abundance that was native rangeland dried to windblown dust. The positive effect of the Dust Bowl era was that it marked the beginning of national programs to conserve the nation's soil and water resources. Credit: USDA Soil Conservation Service.

rangelands and specified the purposes for which those data were to be used: to end soil erosion and to preserve natural resources.[6] Toward this end, SCS field technicians developed a system of range condition classification that field staff and ranchers could understand and use in the development of ranch conservation plans to minimize soil erosion (Helms, 1990). Although these determinations of range condition were a regular responsibility of SCS field technicians, no data from these individual conservation plans were summarized in a report on the conditions of nonfederal rangelands.

CONGRESS MANDATES SURVEY OF WESTERN RANGELANDS

In 1936, Resolution 289 (U.S. Congress, Senate, 1936) requested from USDA a report summarizing information on western rangelands. The Congress justified its request in the resolution by stating that large parts of the western range had been subject to unrestricted use since settlement and were commonly believed to be seriously depleted, that the range resource constituted one of the major sources of wealth to the nation and, that USDA had accumulated a large amount of information on the condition of the range resource, the factors causing degradation, and the social and economic importance of rangelands.

The request for this report points to the lack of a national-level inventory or a mechanism to synthesize data collected by range managers to regulate grazing or evaluate management practices. Although the base data were unpublished, summary tables were presented within the report. Forage depletion by vegetation type was estimated in terms of decline from the original forage values (pristine condition). Results were summarized by the following depletion classes: 0–25, 26–50, 51–75, and 76–100 percent. Remnants of pristine range and protected areas—such as cemeteries and railroad rights of way—were used for comparison. Thus, the report represented the professional opinion of range scientists about the degree of depletion of both federal and nonfederal rangelands.

More Recent Surveys

From 1936 to 1966, range conditions were not surveyed at the national level (Box, 1990). In the early 1960s, the Public Land Law Review Commission was established to study the management of and whether the public lands should be kept in public ownership. As part of their deliberations, they commissioned Pacific Consultants, a private consulting firm, to do a nationwide range survey. Pacific Consultants, however, did not conduct original on-site field surveys. Instead, they collected information from the files of the federal agencies and other sources and made an assessment of U.S. rangelands. In 1972, the USFS made an assessment of the condition of the federal rangelands. It appears the assessment was done by using data already collected by federal agencies and perhaps used previously in the survey by Pacific Consultants (Box, 1990).

Environmental Legislation Increases Information Needs

The wave of federal environmental legislation that commenced with the Wilderness Act of 1964[7] included several laws that contained significant inventory mandates for SCS, USFS, and BLM. Although some of these laws required more detail from agencies with respect to their on-the-ground management, other laws expanded the role of federal agencies in inventories at the national level. During this period, the U.S. Congress also enacted several laws with peripheral, but nonetheless important, relevance to rangeland inventories.

NATIONAL ENVIRONMENTAL POLICY ACT

The first of these laws, the National Environmental Policy Act of 1969,[8] requires all federal agencies to write environmental impact statements on all proposals for major federal actions that significantly affect the human environment. In 1974, a court ordered BLM to prepare such statements for all of its local grazing programs according to an agreed-upon schedule.[9] That ruling was premised on and impelled by the evidence before the court of extensive resource damage on BLM-managed rangelands caused by livestock grazing. The National Environmental Policy Act brought the need for comprehensive inventorying and monitoring of rangelands into stark relief because it is difficult, if not impossible, to assess the environmental consequences of grazing without conducting baseline inventories and incorporating the resultant information into environmental impact statements. Although by 1988 BLM did complete the 144 site-specific grazing environmental impact statements required by law, it did not do all of the necessary site-specific inventories, in part because of budgetary problems. Data on 16 percent of BLM-managed lands were more than a decade old (U.S. General Accounting Office, 1988a). In 1985, an examination of 116 environmental impact statements indicated that BLM had used different terminologies and methodologies to assess rangelands (Wald and Alberswerth, 1985). Such methodological differences make it difficult to compare the results of rangeland assessments across the United States.

CLEAN WATER ACT

The Clean Water Act of 1972,[10] as amended in 1987, requires states to develop and employ the best management practices to control nonpoint sources of water pollution, which include range livestock grazing, silvicultural activities (silviculture is a branch of forestry that deals with the development and care of forests), and other agricultural practices (Thompson, 1989; Whitman, 1989). The Clean Water Act requires each state to adopt water quality standards for its waters. Because federal agencies must comply with those standards,[11] they need to obtain water quality information. The act's requirements have had some impact on timber harvesting on public lands[12] and may have important implications for the kind of information needed from surveys of federal and nonfederal rangelands.

ENDANGERED SPECIES ACT

The Endangered Species Act of 1973[13] requires federal agencies to protect the wildlife species listed in the Act and their habitats in several distinct ways (Coggins and Russell, 1982). Compliance with the Act's requirements and implementing regulations[14] may well necessitate special resource inventories. The Act has also had an effect on timber harvests,[15] but as yet it has not been a major influence on rangeland inventories or monitoring.

Legislation Spurs Development of National Inventories

The U.S. Congress passed a series of laws in the 1970s that redefined the responsibilities of federal agencies to conduct and analyze data on natural resources on both federal and nonfederal lands. The Resources Planning Act, the Forest and Rangeland Renewable Resources Planning Act, and the Soil and Water Resources Conservation Act mandated efforts by USFS, BLM, and SCS to inventory and monitor rangelands.

RESOURCES PLANNING ACT

In 1974 Congress passed the Resources Planning Act of 1974,[16] in which it directed the USFS to conduct and update every 10th year a detailed renewable resource assessment. Each assessment must include the following, among other things: "an inventory, based on information developed by the Forest Service and other Federal agencies, of present and potential renewable resources, and an evaluation of opportunities for improving their yield of tangible goods and services, together with estimates of investment costs and direct and indirect returns to the Federal Government."[17] The assessments and inventories on which they are based are intended to provide information and analysis for use in the development of national-level resource policies and programs as well as site-specific management plans.[18]

The Resources Planning Act specifically directs the USFS to analyze the present and anticipated uses of, demand for, and supply of the nation's renewable resources. Timber, wildlife, range, water, minerals, and lands have been analyzed in the three assessments completed since the Act was passed. The need to collect more data to support the Resources Planning Act has been recognized (McClure et al., 1979) and has led the USFS to expand the timber survey, now called the forest inventory (U.S. Department of Agriculture, U.S. Forest Service, 1985). However, this inventory surveys only pinyon-juniper rangelands (a habitat type dominated by one of several species of pinyon pine [*Pinus* spp.] or juniper [*Juniperus* spp.]) and some shrublands. The 1989 assessment analyzed the federal and nonfederal rangelands from a national perspective, reporting data on the condition of rangelands gathered by USFS, BLM, and SCS (Joyce, 1989).

FOREST AND RANGELAND RENEWABLE RESOURCES PLANNING ACT

Congress passed the Forest and Rangeland Renewable Resources Research Act[19] to complement the policies and direction set forth in the Resources Planning Act. The Act includes a provision intended to "ensure the availability of adequate data and scientific information for development" of the assessment.[20] It requires the USFS to keep current a comprehensive survey and analysis of the present and prospective conditions of and requirements for renewable resources of the forests and rangelands of the United States; the supplies of such renewable resources, including a determination of the present and potential productivity of the land; and any other facts that may be necessary and useful in the determination of ways and means needed to balance the demand for and supply of these renewable resources, benefits, and uses in meeting the needs of the people of the United States.[21]

SOIL AND WATER RESOURCES CONSERVATION ACT

The legislative mandate to SCS for national-level inventories is similar to that of the USFS. In 1977, Congress, recognizing a lack of data necessary to facilitate "informed long-range policy decisions with respect to the conservation and improvement of our country's soil and water resources" enacted the Soil and Water Resources Conservation Act.[22] The Act made the SCS responsible for carrying out a "continuing appraisal of the soil, water, and related resources" of the United States.[23] The phrase "soil, water, and related resources" is defined broadly as "those resources which come within the scope of the programs administered and participated in by the Secretary of Agriculture through the Soil Conservation Service."[24]

In the Soil and Water Resources Conservation Act, Congress declares that "[r]esource appraisal is basic to effective soil and water conservation" and that a coordinated appraisal program is "essential" since decisions affecting soil and water resources are made by a variety of individuals and administrative agencies and may affect other decisions and programs.[25] The appraisal "shall" include

1. data on the quality and quantity of soil, water, and related resources, including fish and wildlife habitat;

2. the capability and limitations of those resources for meeting current and projected demands on the resource base; [and]

3. the changes that have occurred in the status and condition of those resources resulting from various past uses, including the impact of farming technologies, techniques, and practices.[26]

Regarding the sources of information to be used in the appraisal, the Act states that the SCS shall use not only data collected under the Soil and Water Resources Conservation Act but also "pertinent data and information collected by the Department of Agriculture and other Federal, State, and local agencies and organizations."[27]

NATIONAL RESOURCES INVENTORY

The appraisal mandated by the Soil and Water Resources Conservation Act, now termed the National Resources Inventory, must be carried out at periodic intervals. The next of these is due by December 31, 1995.[28] The secretary of USDA may require that interim appraisals be made. A description of the current National Resources Inventory is given later in this chapter.

There have been two appraisals since passage of the Soil and Water Resources Conservation Act in 1977. Range condition measures have been reported for nonfederal U.S. rangelands as part of both appraisals. The most recent assessment estimated the areas of nonfederal rangelands that are in excellent, good, fair, and poor range condition on the basis of inventory data from 1982 (U.S. Department of Agriculture, Soil Conservation Service, 1989a).

ENVIRONMENTAL MONITORING AND ASSESSMENT PROGRAM

In 1988, the U.S. Environmental Protection Agency (EPA) developed the Environmental Monitoring and Assessment Program (EMAP) in response to a recommendation from the EPA Science Advisory Board that such a program be implemented to monitor ecological status and trends that would identify emerging environmental problems before they reach crisis proportions (U.S. Environmental Protection Agency, 1992). EMAP will coordinate the research and the monitoring and assessment efforts needed to both document the current condition of ecological resources and predict the effect of different management alternatives on those resources.

The EMAP program has three broad objectives: (1) estimate the current status, extent, changes, and trends in indicators of the condition of U.S. ecological resources; (2) monitor indicators of pollutant exposure and habitat condition and seek associations between human-induced stresses and ecological condition; and (3) provide periodic statistical summaries and interpretive reports on ecological status and trends to resource managers and the public (U.S. Environmental Protection Agency, 1992). EMAP will coordinate several long-term monitoring efforts to collect data at regional scales from eight resource categories: arid lands, agricultural systems, forests, lakes and streams, the Great Lakes, inland and coastal wetlands, estuaries, and coastal waters. The monitoring program includes collection and interpretation of field data as well as statistical analysis and sampling design and development of ecological indicators. The arid lands program, which includes rangelands, is currently undertaking research to develop ecological indicators and rangeland classification systems and to explore the use of remote sensing to monitor changes in arid lands.

Inventory Mandates for Land Management and Planning

Both the USFS and BLM have been given inventory responsibilities in connection with their land use planning and management duties. Three major pieces of legislation, two enacted in 1976 and one enacted in 1978, provide the framework for management, inventory, and planning for both agencies.

FEDERAL LAND POLICY AND MANAGEMENT ACT

The Federal Land Policy and Management Act of 1976[29] is called BLM's "Organic Act," but sections 1751 through 1753 deal with range management by both BLM and USFS. In essence, these sections declare that "a substantial amount of the Federal range lands is deteriorating in quality,"[30] call for improvement in current conditions,[31] provide funds for "range improvements,"[32] authorize adjustments to grazing privileges depending on the condition of the rangeland,[33] and sanction other management actions.

Concerning inventories of rangelands, Congress stated: "The [BLM] shall prepare and maintain on a continuing basis an inventory of all public lands and their resource and other values . . ., giving priority to areas of critical environmental concern."[34] That inventory is to form the basis of land use plans that are required to produce multiple-use, sustained-yield management of the surface resources of BLM-managed public lands.[35] Livestock management of these lands is but a subset of multiple-use management,[36] which the Federal Land Policy and Management Act defines to mean management to produce a harmonious and optimum combination of outdoor recreation, range, wildlife, watershed, and timber production. The Act specifies that economic optimization is not the goal.[37]

SOIL-VEGETATION INVENTORY METHOD

In 1977, BLM adopted the site inventory method, which was later expanded into the soil-vegetation inventory method, to standardize rangeland sampling for environmental impact statement and allotment management plan reporting. Within this inventory, soil maps, site descriptions, and aerial photographs were used to classify rangelands for survey purposes (Wagner, 1989). Ground checks determined range site, woodland type, or forest type, as well as soil type. Sampling units called site write-up areas were defined on the basis of the vegetation communities that were present on the site. Data on basal and canopy ground covers and vegetative production were collected on site write-up areas and combined with data from additional studies on phenology (a branch of science that deals with the relations between climate and periodic biological phenomena), climate, actual use, and utilization (Wagner, 1989).

The soil-vegetation inventory method was intended to be the official inventory method for basic inventories of soil and vegetation by BLM, but it was not to preclude the use of site-specific studies for special purposes. The soil-vegetation inventory method, however, was discontinued in the early 1980s and never became the official standardized inventory system for federal rangelands managed by BLM.

NATIONAL FOREST MANAGEMENT ACT

Like the BLM, the USFS is also required to manage livestock within the context of multiple uses and sustained yields on the lands within its jurisdiction.[38] The National Forest Management Act of 1976[39] directs USFS to "develop and maintain on a continuing basis a comprehensive and appropriately detailed inventory of all National Forest System lands and renewable resources."[40] This inventory, like the BLM inventory, is to be kept current "so as to reflect changes in conditions and identify new and emerging resources and values."[41] As is the case with BLM, the inventory information is to be used in the development of land use plans.

Within the national forest system, inventories exist for range; timber;

Four bull elk with fully grown antlers but still in the velvet. These elk are on a summer range in Montana. Credit: USDA U.S. Forest Service.

soils and geology; natural water occurrences, including quality and quantity and wetlands and floodplains; existing plant life, including threatened and endangered species; existing fish and wildlife, including threatened and endangered species; habitat conditions for selected vertebrate or invertebrate species; and quantitative data for determining species and community diversity (U.S. Department of Agriculture, U.S. Forest Service, 1985). The earliest rangeland inventory was done on the Coconino National Forest in 1911 in response to a question regarding the livestock carrying capacity of the rangeland (U.S. Department of Agriculture, U.S. Forest Service, 1985). Methodologies for inventorying rangelands have varied since then, but they have primarily focused on the vegetation within allotments. Information needs have included determination of the suitability of lands for livestock grazing, the kinds of plant communities present, the ecological status of those communities, livestock forage conditions, the status of soil cover and soil stability, the trends in ecological status and soil stability, the capability of the lands to produce suitable food and cover for wildlife, and the effects of grazing. Current inventory procedures are described later in this chapter. Development of specific stands and guidelines for range inventory are the responsibility of the regional forester; thus, variations between regions are permitted.

PUBLIC RANGELANDS IMPROVEMENT ACT

The Public Rangelands Improvement Act of 1978,[42] which is also applicable to both BLM and USFS, directly addresses the issue of range condition measurement.[43] The act's policies include a federal commitment to "inventory and identify current public rangeland conditions and trends as part of the inventory process required by [the Federal Land Policy and Management Act]," and to "manage, maintain, and improve the condition of the public rangelands so that they become as productive as feasible for all rangeland values in accordance with management objectives and the land use planning process."[44] Congress thereafter defined its terms as follows[45]:

> The term "range condition" means the quality of the land reflected in its ability in specific vegetative areas to support various levels of productivity in accordance with range management objectives and the land use planning process, and relates to soil quality, forage values (whether seasonal or year round), wildlife habitat, watershed and plant communities, the present state of vegetation for that site, and the relative degree to which the kinds, proportions, and amounts of vegetation in a plant community resemble that of the desired community for that site.
>
> The term "native vegetation" means those plant species, communities, or vegetative associations which are endemic to a given area and which

would normally be identified with a healthy and productive range condition occurring as a result of the natural vegetative process of the area.[46]

Carrying out one of its policies, Congress said that both agencies shall update, develop, and maintain on a continuing basis . . . an inventory of range conditions on the public rangelands, and shall categorize or identify such lands on the basis of the range conditions and trends thereof as they deem appropriate. Such inventories . . . shall be kept current on a regular basis so as to reflect changes in range conditions; and shall be available to the public.[47]

The Act directs that when grazing is allowed on BLM-administered public lands, the management goal "shall be to improve the range conditions of the public lands so that they become as productive as feasible for all rangeland values."[48]

Statutory Requirements for Inventories and Monitoring

The SCS, USFS, and BLM are required to inventory rangelands for all resources, albeit for a variety of purposes. The laws cited in the previous sections mandate, request, or imply the need for significant amounts of information for managing, planning, and inventorying rangelands. An assessment of rangelands is a statutorily essential component of the inventories of all three agencies, and all three agencies are specifically required to consider watershed, recreation, and wildlife as part of either their inventory responsibilities (SCS) or other duties (SCS, USFS, and BLM). The purpose of inventories is to provide the necessary planning and management information. Recent legislation has increased the breadth and depth of the information required from inventories. Congress has left federal agencies considerable discretion to devise and adopt standards and methods to inventory and monitor rangelands. Congress has made clear, however, that rangeland inventory and monitoring systems should

• cover all rangeland resources, specifically including watersheds, and should emphasize determination of the quality of the land;
• relate to the congressional mandate to achieve sustained yields or health of all renewable resources;
• lead to improvement in the status of rangelands; and
• be useful for planning and management purposes as well as for purely informational purposes.

CURRENT INVENTORYING AND MONITORING SYSTEMS

There are great differences between the inventory and monitoring systems currently used on nonfederal versus federal rangelands.

Nonfederal Rangelands

SCS surveys all nonfederal lands every 5 years as part of the National Resources Inventory, which is mandated by the Soil and Water Resources Conservation Act of 1977. In 1987, nearly 163 million hectares (402 million acres) of nonfederal lands, excluding Alaska, were classified as rangelands. The information gathered through the National Resources Inventory process is used in the Soil and Water Resources Conservation Act appraisal published by SCS every 10 years. The data used in the National Resources Inventory are collected by census, area sampling, and point sampling methods (U.S. Department of Agriculture, Soil Conservation Service, 1989b). (Area sampling involves stratification of the land base and selection of a primary sampling units from each stratum. Point sampling is the collection of data such as plant composition, slope, and soil type at points designed within each primary sampling unit.) Census data are used to identify nonfederal lands, which are the lands the SCS then inventories for the National Resources Inventory using area and point sampling methods. Major land resource areas were the basis for stratification in the 1977, 1982, and 1987 National Resources Inventories; and the 1982 inventory was statistically reliable to that level.

For the United States as a whole, nearly 350,000 permanent primary sampling units have been established, and a random selection is sampled for each National Resources Inventory. Each of the three National Resources Inventories conducted by SCS has used a different number of primary units: 77,000 sites in the 1977 inventory, more than 300,000 in the 1982 inventory, and 100,000 in the 1987 inventory (U.S. Department of Agriculture, Soil Conservation Service, 1987).

Primary sampling units come in four sizes: 16, 40, 65, and 299 hectares (40, 100, 160, and 640 acres, respectively). For each primary sampling unit there are computer-selected points from which data are gathered. For example, there are two such points on 16-hectare (40-acre) primary sampling units and three on 65-hectare (160-acre) primary sampling units. Data are collected over the entire primary sampling unit (area sampling) and at each designated point in the primary sampling unit (point sampling). Both kinds of data are then expanded to produce totals for a state.

Instructions for the 1987 National Resources Inventory required each randomly selected primary sampling unit to be located on an aerial photograph or map. They directed that specified data—those for urban and built up lands, farmsteads, critically eroding areas, and water bodies—within the sampling area be delineated from photographs on a preprinted work sheet developed specifically for the primary sampling unit and that inventory (U.S. Department of Agriculture, Soil Conservation Service, 1987). Because the 1987 survey was designed to update the 1982 National

Resources Inventory (U.S. Department of Agriculture, Soil Conservation Service, 1987), the work sheet included data obtained during the previous effort(s). The instructions required data from previous inventories to be verified and revised and identified the new data that were to be obtained. The instructions prescribed how each item was to be addressed, including the way that these measurements were to be validated in the field.

The instructions for point sampling required the collection of more detailed information for each primary sampling unit—range site (SCS) and range condition (SCS), apparent trend (SCS), kinds of crops, types of conservation practices in use and needed, soil characteristics, erosion, and wetlands—either during actual visits to established points or through the use of maps, photographs, or remote sensing. They called for range condition (SCS) to be determined by the SCS methods described in Chapter 3 of this report.

The systematic collection of range condition (SCS) data as part of the National Resources Inventory allows SCS to make estimates of the area of nonfederal rangelands in excellent, good, fair, or poor range condition (SCS) as determined by their resemblance to the defined climax plant community (SCS) thought to be characteristic of that site. The National Resources Inventory system permits national-level assessments of range condition (SCS) on nonfederal rangelands. Data are collected by a statistically valid sampling scheme and by using standardized definitions and methodologies. The assessment of range condition (SCS) that is part of this system is not an adequate assessment of rangeland health, but the system to collect, analyze, and aggregate data that could be used to assess the health of nonfederal rangelands is in place.

Federal Rangelands

There is no statistically designed survey of federal rangelands comparable to the National Resources Inventory of nonfederal rangelands. Each agency responsible for managing federal rangelands collects data on the rangelands under its jurisdiction using the methods selected by each agency.

BUREAU OF LAND MANAGEMENT

BLM has responsibility for managing 109 million hectares (270 million acres) of land, approximately 69 million hectares (170 million acres) of which have been classified as rangeland (U.S. Department of the Interior, Bureau of Land Management, 1989). BLM recognizes that monitoring and evaluation are essential management functions in (1) establishing and evaluating progress in meeting resource management objectives, (2) developing management plans, (3) preparing environmental analyses, and

(4) supporting decision making (U.S. Department of the Interior, Bureau of Land Management, 1985a:Section 4400.06). BLM's national policy calls for the rangelands under its jurisdiction to be inventoried and classified according to their ecological status (USFS and BLM). However, BLM does not gather data primarily for a national inventory. Most of the available BLM rangeland data are used to manage individual grazing allotments. These data are not collected by a statistically designed sampling method that would allow confident aggregation of results on a national basis. BLM's ecological status (USFS and BLM) and trend data have, however, been combined in the past and have been used in national reports, including the assessment required by the Resource Planning Act of 1974.

U.S. FOREST SERVICE

USFS has responsibility for managing 74 million hectares (182 million acres) of land, of which about 16 million hectares (40 million acres) are classified as rangelands (U.S. Department of Agriculture, U.S. Forest Service, 1989b). Management plans are developed by USFS personnel for units of land such as livestock grazing allotments, timber management units, or watersheds. All of the units and management activities are combined into forest plans for each national forest.

Standards for ecosystem classification and interpretation for use by all units of the national forest system have been recently adopted (U.S. Department of Agriculture, U.S. Forest Service, 1991b). Regional offices will continue to have the flexibility they need to select the methods that they consider appropriate to meet the standards. The new policy states that USFS shall use ecological type classification to "coordinate and integrate resource inventories to stratify land and resource production capability and make predictions and interpretations for management" (U.S. Department of Agriculture, U.S. Forest Service, 1991c:Section 2060.3). The purposes of this new policy are (1) to provide an integrated ecosystem classification based on potential natural community, soils, and physical characteristics; (2) to provide a unifying ecosystem framework for use in land and resource management;" and (3) to develop an ecologically based information system to aid in evaluating land capability, interpreting ecological relationships, and improving multiple use management (U.S. Department of Agriculture, U.S. Forest Service, 1991c:Section 2060.2).

In addition to utilizing information regarding ecological type (USFS) and ecological status (USFS and BLM), the USFS system also includes resource value ratings for livestock, wildlife, and other uses. These ratings are intended to permit the agency to assess the degree to which the vegetation on a site satisfies or meets previously established vegetation management objectives (Society for Range Management, 1989:5).

Current USFS inventory procedures, however, do not produce statistically reliable estimates of the proportion of rangelands in each ecological status (USFS and BLM) class. Different methods are used to measure ecological status in different USFS regions. In addition, the ecological status (USFS and BLM) data are collected as part of ongoing USFS management activities rather than as part of a representative sampling program. The differences in methods used and the absence of a statistically reliable sampling design do not allow confident compilation of USFS ecological status (USFS and BLM) data at the national level.

Current Inventories Are Inadequate

Recent U.S. General Accounting Office reports suggest that both BLM and USFS are limited in their ability to obtain adequate inventory information. About two-thirds of BLM allotments and one-fourth of USFS allotments did not have management plans, and data for another 16 percent of BLM-managed rangelands and 31 percent of USFS-managed rangelands were more than a decade old (U.S. General Accounting Office, 1988a). The USFS's forest planning process rests on the availability of information on all resources, and insufficient or inadequate data hamper this process (U.S. Department of Agriculture, U.S. Forest Service, 1990a). Fosburgh (1986) noted the lack of an adequate information flow between forest planning processes and the national-level planning process as envisioned in the Resources Planning Act and suggested that this was a failure of the Act.

Professional organizations such as the Society for Range Management (1989) have tried to bring the various reports of the SCS, BLM, and USFS together into a single document. Unfortunately, the data for those reports were originally gathered for different purposes, at different times, by using different techniques and formats. In this example, USFS reports ecological status on rangelands, whereas SCS and BLM report range condition.

Conservation groups such as the National Wildlife Federation and the Natural Resources Defense Council have examined data reported by BLM for the portion of U.S. rangelands under its jurisdiction (Wald and Alberswerth, 1989). Oversight agencies such as the U.S. General Accounting Office (1988a,b) have made their own assessments. The U.S. General Accounting Office assessments are in response to House and Senate requests for information. For the U.S. General Accounting Office study on the management of grazing allotments (U.S. General Accounting Office, 1988a), a questionnaire was used to obtain data from BLM and USFS because it was considered impractical to conduct on-site visits at more than a few field offices.

Some individuals (Box, 1990; Box et al., 1976) have applied their professional judgments to the various historical surveys to try to assess historical changes in rangelands. These data sources range from documents such as the 1936 Senate report (U.S. Congress, Senate, 1936) to compilations of data from different agencies, and they cannot be considered adequate surveys of the nation's rangelands. Current rangeland inventories simply do not provide the data needed to support national assessments of rangeland health.

NATIONAL SYSTEM OF INVENTORYING AND MONITORING RANGELAND HEALTH IS NEEDED

The structure of an inventory involves consideration of its appropriate geographical scope, the intensity of sampling within that geographical scope, the kind of land to be sampled, and the attributes to be sampled. The questions that are asked at the national level require a different level of sampling intensity than the questions that are asked about a specific allotment on USFS- or BLM-administered lands. The questions asked about rangelands involve both rangeland monitoring and rangeland inventorying.

Inventories establish the status of a rangeland's resources at a given point in time. Resources are evaluated on the basis of either direct measurement or statistical inference taken from sampling resources indicators. Monitoring measures changes in the status of selected resources or indicators over time. It may involve a complete reinventory, but it more commonly involves repeated measurements using the same methods on selected areas that represent larger areas (Society for Range Management, Range Inventory Standardization Committee, 1983). Monitoring measures selected attributes on rangelands that can be accurately remeasured to determine changes in those attributes. The attributes selected (that is, vegetation, soil, or animals) determine the environmental changes that can be detected. Although rangeland inventories have been implemented, few would qualify as rangeland monitoring.

A national-level assessment of rangeland health requires the following:

- adoption of a standardized and consistent definition of rangeland health and of measurable indicators of change in rangeland health;
- consistent and well-correlated classification of federal and nonfederal rangelands;
- collection of data by the same or similar methods that will enable the data to be combined on a national level;
- collection of data on the basis of a statistically valid sampling

scheme that enables data to be evaluated at the state, regional, and national levels; and

• periodic and consistent repetition of sampling to detect trends in the measures used to evaluate rangeland health.

A system that can be used to produce statistically reliable estimates of the health of rangelands is not in place for either federal or nonfederal rangelands. There is an urgent need to develop a system to inventory rangeland health to judge whether current management and use of the nation's federal and nonfederal rangelands are adequately conserving the capacity of rangelands to produce commodities and satisfy values. The fundamental output of a national monitoring system should be the collection and reporting of a set of data of measurable indicators of rangeland health. These data can be used to estimate the proportion and distribution of federal and nonfederal rangelands that are healthy, at risk, or unhealthy and to determine whether current use and management are conserving the productive capacity of the nation's rangelands.

Convene an Interagency Task Force

The secretaries of USDA and DOI should convene an interagency task force to develop, test, and standardize indicators and methods for inventorying and monitoring rangeland health on federal and nonfederal rangelands.

Standardized indicators are needed for each of the criteria recommended in Chapter 4: degree of soil movement by wind and water, distribution of nutrients and energy in space and time, and plant demographics. The indicators suggested in Chapter 4 should serve as a useful starting point in the development of standardized indicators of rangeland health, but they must be refined and tested.

Standardized methods to measure indicators and to classify rangelands as healthy, at risk, and unhealthy are also needed. The preliminary decision rules described in Chapter 4 (see Table 4-7) should serve as a useful starting point for determining when a rangeland is healthy, at risk, or unhealthy, but these classification decisions must be refined and tested.

The multiagency task force should coordinate federal efforts, including EPA's EMAP, leading to

• a set of indicators that should be included in a minimum data set for inventorying and monitoring rangeland health,

• standard methods of measuring indicators of rangeland health,

• a series of field tests to validate the indicators and methods selected, and

• quantification of the correlation between measures of rangeland health and range condition (SCS) or ecological status (USFS and BLM).

National Sampling System Needed

The secretaries of USDA and DOI should develop coordinated plans for implementing a sampling system on federal and nonfederal rangelands that will produce estimates of the proportion of healthy, at-risk, and unhealthy rangelands that are significant at an appropriate substate level.

A national sampling system that coordinates the activities of the USDA, DOI, and EPA is needed to collect, analyze, and aggregate data to determine the proportion of federal and nonfederal rangelands that are healthy, at risk, or unhealthy. The National Resources Inventory, conducted by SCS, provides a statistically valid sampling design for nonfederal rangelands. The addition of standardized indicators of rangeland health to the National Resources Inventory can produce statistically valid estimations of the proportions of nonfederal rangelands that are healthy, at risk, or unhealthy.

No comparable sampling program is in place on federal rangelands. Most of the data collected are for management purposes rather than for inventorying and monitoring purposes. The development of a coordinated sampling system for both federal and nonfederal rangelands is urgently needed.

Periodic Sampling Needed

The secretaries of USDA and DOI should develop coordinated plans for implementing periodic sampling of federal and nonfederal rangelands to determine changes in the proportions of healthy, at-risk, and unhealthy rangelands.

Periodic monitoring must be a fundamental part of a valid national system for evaluating rangeland health. The periodicity of repeated sampling should reflect the rapidity of change within the indicators selected to monitor rangelands and the degree of degradation that a change implies to give adequate early warning of increases in the area of unhealthy rangelands. Monitoring should be periodic enough such that a rangeland would not slip from a healthy to an unhealthy state between sampling periods.

Transition to Rangeland Health

Indicators of soil surface condition should be added to all current and ongoing range condition (SCS) and ecological status (USFS and BLM) assessments, and any other ongoing efforts to assess rangelands, as a first step toward a more comprehensive evaluation of rangeland health.

There is much experience with the use of soil surface characteristics as indicators of soil stability and watershed function. The addition of indicators of soil surface condition to all current and ongoing efforts to assess rangelands would be a useful first step toward a more comprehensive system of evaluating rangeland health—a step that should be taken immediately. Data on the soil surface condition of rangelands should also be collected as part of the National Resources Inventory.

All current and ongoing rangeland assessments done as part of Resources Conservation Act (RCA) appraisals, Resources Planning Act (RPA) assessments, national forest planning, USFS and BLM land use and allotment planning, and environmental assessments should be based on the analysis of multiple ecological attributes.

SCS, USFS, and BLM should analyze multiple ecological attributes of rangelands as part of current rangeland assessments and appraisals. Currently, plant composition and, in some cases, biomass production are the only attributes systematically used in rangeland assessments. These data alone are not sufficient for assessing rangeland health. Plant composition and production data collected as part of range condition (SCS) and ecological status (USFS and BLM) ratings should be analyzed in conjunction with information collected on indicators of soil surface condition as recommended above and all other available information on erosion rates. Using these multiple indicators, the agencies could begin to assess soil stability and watershed function, distribution of nutrients and energy, and presence of functioning recovery mechanisms as a means of identifying rangelands at greater risk of loss of health. This analysis should be part of conservation planning or management of grazing allotments as well as national appraisals and assessments.

These assessments would not provide a complete assessment of rangeland health, but they would represent progress toward measuring and analyzing multiple ecological attributes within each agency. They would also help guide national policy for managing federal and nonfederal rangelands in the interim while more comprehensive and systematic assessments of rangeland health are developed.

Basic data on soil surface conditions, erosion rates, plant composition, and biomass production assembled and used to assess rangelands as part of RCA appraisals, RPA assessments national forest planning, environmental assessments, and other assessments of federal and nonfederal rangelands should be made available to the public and the scientific community for independent review.

Independent review will increase the understanding of and confidence in the results of the assessments of federal and nonfederal rangelands. Publication of basic data will provide a data set for scientific evaluation of the utility of alternative indicators of soil stability and watershed function, distribution of nutrients and energy, and presence of recovery mechanisms as measures of rangeland health. The availability

Sheep graze on an area that was clear cut of lodgepole pine for pulp—an experimient in multiple use, grazing during natural regeneration of the forest. Credit: USDA Soil Conservation Service

of basic data, for example, used to estimate erosion rates as part of the National Resources Inventory has allowed scientists to test the effect of alternative agricultural policies and crop management practices on erosion rates. Independent review of these basic data has increased the confidence of estimates of erosion reductions expected from changes in farming practices. It is important that basic data on multiple ecological attributes of federal and nonfederal rangelands be made available to both the public and the scientific community to accelerate the transition to comprehensive methods for assessing rangelands.

SCS, USFS, and BLM should continue current and ongoing range condition (SCS) and ecological status (USFS and BLM) ratings while the transition to rangeland health assessment is made.
The data that have been and continue to be collected for range condition (SCS) and ecological status (USFS and BLM) assessments should provide a critical historical data set for use in judging changes in rangeland conditions. As a transition is made to national-level inventorying and monitoring of rangeland health as recommended here, it is imperative that this information not be lost. The committee strongly recommends that current and planned monitoring efforts that use range condition (SCS) and ecological status (USFS and BLM) move ahead and be augmented by the collection of additional data for evaluating rangeland health.

NOTES

1. 16 USC § 471 (1891) (repealed).
2. 16 USC § 475 (1897).
3. 43 USC § 315 and 315(m) (1934).
4. *Ibid.* preamble.
5. 16 USC § 590(a) to 590(q).
6. 16 USC § 590(a).
7. 16 USC § 1131 to 1136.
8. 16 USC § 4221 to 4247.
9. *Natural Resources Defense Council* v. *Morton,* 388 F. Supp. 829 (D.D.C. 1974), *aff'd,* 527 F.2d 1386 (D.C. Cir. 1976), *cert. denied* 427 U.S. 913 (1976).
10. 33 USC § 1251 to 1376 (1977).
11. *Northwest Indian Cemetery Protective Ass'n* v. *Peterson,* 795 F.2d. 688 (9th Cir. 1986, *rev'd on other grounds* 485 U.S. 439 (1988).
12. *Id.*
13. 16 USC § 1531 to 1543 (1973).
14. 50 CFR § 402 (1992).
15. *Sierra Club* v. *Lyng,* 694 F. Supp. 1260 (E.D.Tex. 1988).
16. 16 USC § 1600 (1974) *et seq.*
17. 16 USC § 1601(a)(2) (1974).
18. *See, e.g.,* 16 USC § 1601(a)(4) (1974).
19. 16 USC § 1641(b) (1978).
20. 16 USC § 1642(b) (1978).
21. *Id.*
22. 16 USC § 2001 (1977) *et seq.*
23. 16 USC § 2004(a) (1977).
24. 16 USC § 2002(2) (1977).
25. 16 USC § 2001(3) (1977).
26. 16 USC § 2004(a)(1) to 2004(a)(3) (1977).
27. 16 USC § 2004(b) (1977).
28. 16 USC § 2004 (1977).
29. 43 USC § 1701 to 1784 (1976).
30. 43 USC § 1751(b)(1) (1976).

31. *Id.*
32. *Id.*
33. 43 USC § 1752(e) (1976).
34. 43 USC § 1711(a) (1976).
35. 43 USC § 1712(c)(1) (1976).
36. 43 USC § 1732(a) (1976).
37. 43 USC § 1702(h) (1976).
38. 16 USC § 528(a), 531(b) (1960).
39. 16 USC §§ 472(a), 576(b), 1611 to 1614 (1976).
40. 16 USC § 1603 (1976).
41. *Id.* See also 43 USC § 1711(a) (1976).
42. 43 USC § 1901 to 1908 (1978).
43. 43 USC § 1901 (1978).
44. 43 USC § 1901(b) (1978).
45. 43 USC § 1902 (1978).
46. 43 USC § 1902(d) and 1902(e) (1978).
47. 43 USC § 1903(a) (1978).
48. 43 USC § 1903(b) (1978).

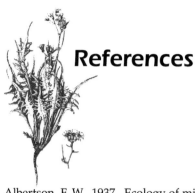

References

Albertson, F. W. 1937. Ecology of mixed prairie in West Central Kansas. Ecological Monographs 7:481–547.

Archer, S. 1989. Have southern Texas savannas been converted to woodlands in recent history? The American Naturalist 134:545–561.

Athearn, R. G. 1986. The Mythic West in Twentieth Century America. Lawrence: University of Kansas Press.

Barmore, R. L. 1980. Soil Survey of Yuma-Wellington Area, Arizona and California. Arizona Agriculture Experiment Station and California Agriculture Experiment Station. Washington, D.C.: Soil Conservation Service, U.S. Department of Agriculture.

Beetle, A. A. (compiler). 1954. Introductory bibliographies in range management and plant ecology. II. Range management literature 1890–1905. Wyoming Range Management 67:1–10.

Bentley, J. R., and M. W. Talbot. 1951. Efficient use of annual plants on cattle ranges in the California foothills. Pp. 1–52 in Circular No. 870. Davis: University of California.

Birkeland, P. W. 1984. Soils and Geomorphology. New York: Oxford University Press.

Blaisdell, J. P., and R. C. Holmgren. 1984. Managing Intermountain Rangelands— Salt Desert Shrub Ranges. General Technical Report INT-163:52. Ogden, Utah: Intermountain Forest and Range Experiment Station.

Bormann, F. H., and G. E. Likens. 1979. Catastrophic disturbance and the steady state in northern hardwood forests. American Scientist 67:660–669.

Box, T. W. 1990. Rangelands. Pp. 101–120 in Natural Resources for the 21st Century, R. N. Sampson and D. Hair, eds. Washington, D.C.: Island Press.

Box, T. W., D. D. Dwyer, and F. H. Wagner. 1976. Condition of the western rangelands. Unpublished report. Council on Environmental Quality, Washington, D.C.

Brady, N. C. 1990. The Nature and Properties of Soils, 10th Ed. New York: Macmillan.

Branson, F. A., G. F. Gifford, K. G. Renard, and R. F. Hadley. 1981. Rangeland Hydrology, 2nd Ed. Denver: Society for Range Management.

Tapertip hawksbeard (*Crepis Acuminata*)

Brewer, R. 1989. The Science of Ecology. Philadelphia: Saunders College Publishing.

Brooks, R. T. 1992. The New England Forest: Baseline for New England Forest Health Monitoring. Resource Bulletin No. 124. Radnor, Penn.: Northeastern Forest Experiment Station, U.S. Department of Agriculture, Forest Service.

Burke, I. C., C. M. Yonker, W. J. Parton, C. V. Cole, K. Flash, and D. S. Schimel. 1989. Texture, climate and cultivation effects on soil organic matter content in U.S. grassland soils. Soil Science Society of America Journal 53:800–805.

Burcham, L. T. 1961. Cattle and range forage in California: 1770–1880. Pp. 140–150 in Agricultural History. The Agricultural History Society. Berkeley: University of California Press.

Burkman, W. G., and G. D. Hertel. 1992. Forest health monitoring: A national program to detect, evaluate, and understand change. Journal of Forestry 90:27.

Chaney, E., W. Elmore, and W. S. Platts. 1990. Livestock Grazing on Western Riparian Areas. Eagle, Idaho: Northwest Resource Information Center, Inc.

Chapline, W. R., and R. S. Campbell. 1944. History of western range research. Agricultural History 18(3):127–143.

Christensen, N. L., J. K. Agee, P. F. Brussard, J. Hughes, D. H. Knight, G. W. Minshall, J. M. Peek, S. J. Pyne, F. J. Swanson, S. Wells, J. W. Thomas, S. E. Williams, and H. A. Wright. 1989. Interpreting the Yellowstone Fires. BioScience 39:678–685.

Clements, F. E. 1916. Plant Succession: An Analysis of the Development of Vegetation. Publication 242. Washington, D.C.: Carnegie Institution of Washington.

Coggins, G. C., and I. S. Russell. 1982. Beyond shooting snail darters in pork barrels. Endangered species and land use in America. Georgetown Law Journal 70:1433.

Comptroller General. 1977. Public Rangeland Continue to Deteriorate. U.S. General Accounting Office Report CED-77-88. Washington, D.C.: U.S. General Accounting Office.

Commission on the European Communities, Directorate General for Agriculture. 1990. European Community Forest Health Report, 1989. Luxembourg: Office for Official Publications of the European Communities.

Connell, J. H., and R. D. Slatyer. 1977. Mechanisms of succession in natural communities and their role in community stability and organization. The American Naturalist 111:1119–1144.

Cook, C. W., and J. Stubbendieck. 1986. Range Research: Basic Problems and Techniques. Denver: Society for Range Management.

Cook, R. J., and R. J. Veseth. 1991. Wheat Health Management. St. Paul, Minn.: APS Press.

Cordell, K. 1989. An analysis of the outdoor recreation and wilderness resource situation in the United States: 1989–2040. General Technical Report RM-89. Fort Collins, Colo.: Rocky Mountain Forest and Experiment Station, U.S. Forest Service, U.S. Department of Agriculture.

Croft, A. R., L. Woodward, and D. A. Anderson. 1943. Measurement of accelerated erosion on range-watershed land. Journal of Forestry 41:112–116.

Dormaar, J. F., and W. D. Williams. 1990. Sustainable production from the rough fescue prairie. Journal of Soil and Water Conservation 45:137–140.

Dormaar, J. F., A. Johnston, and S. Smoliak. 1978. Long-term soil changes associated with seeded stands of wheatgrass in southeastern Alberta, Canada. Pp. 623–625 in Proceedings of the 1st International Rangeland Congress. Denver: Society for Range Management.

Drury, W. H., and I. C. T. Nisbet. 1973. Succession. Journal of the Arnold Arboretum 54:331–368.

Dyksterhuis, E. J. 1949. Condition and management of range land based on quantitative ecology. Journal of Range Management 2:104–115.

Dyksterhuis, E. J. 1958. Range conservation as based on sites and condition classes. Journal of Soil and Water Conservation 13:151–155.

Ellison, L. 1949. The ecological basis for judging condition and trend on mountain range land. Journal of Forestry 47:787–795.

Fitter, A. H., and R. K. M. Hay. 1987. Environmental Physiology of Plants, 2nd Ed. London: Academic Press.

Flather, C. H., and T. W. Hoekstra. 1989. An analysis of the wildlife and fish situation in the United States: 1989–2040. General Technical Report RM-178. Fort Collins, Colo.: Rocky Mountain Forest and Experiment Station, U.S. Forest Service, U.S. Department of Agriculture.

Fosburgh, W. 1986. Wildlife issues in the National Forest System. Pp. 159–173 in Audubon Wildlife Report 1986, A. S. Eno, R. L. Di Silvestro, and W. J. Chandler, eds. New York: National Audubon Society.

Foth, H. O. 1990. Fundamentals of Soil Science, 8th Ed. New York: Wiley.

Friedel, M. H. 1991. Range condition assessment and concept of thresholds: A viewpoint. Journal of Range Management 44:422–426.

Friedel, M. H., B. D. Foran, and D. M. Stafford-Smith. 1990. Where the creeks run dry or ten feet high: Pastoral management in arid Australia. Proceedings of the Ecological Society of Australia 16:185–194.

Gee, K. G., L. A. Joyce, and A. G. Madsen. 1992. Factors Affecting the Demand for Grazed Forage in the United States. General Technical Report RM-210. Fort Collins, Colo.: Rocky Mountain Forest and Range Experiment Station, U.S. Forest Service, U.S. Department of Agriculture.

Gleason, H. A. 1926. The individualistic concept of the plant association. Bulletin Torrey Botanical Club 53(1):7–26.

Gray, G., and L. Clark. 1992. The changing science of forest health. American Forests 89:13–16.

Gutierrez, L. T., and W. R. Fey. 1975. Simulation of secondary autogenic succession in the shortgrass prairie ecosystem. Simulation 24:113–135.

Hanson, C. L., and W. C. Whitman. 1938. Characteristics of major grassland types in western North Dakota. Ecological Monographs 8(1):59–114.

Hanson, H., L. D. Love, and M. S. Morris. 1931. Effects of Different Systems of Grazing by Cattle upon a Western Wheat-Grass Type of Range. Colorado Experiment Station Bulletin 377. Fort Collins: Colorado State University.

Harper, J. L. 1977. Population Biology of Plants. London: Academic Press.

Haskell, B. D., B. G. Northon, and R. Costanza. 1993. What is ecosystem health and why should we worry about it? Pp. 3–20 in Ecosystem Health: New

Goals for Environmental Management, R. Costanza, B. G. Norton, and B. D. Haskell, eds. Washington, D.C.: Island Press.

Heady, H. F. 1975. Rangeland Management. New York: MacGraw-Hill.

Helms, D. 1990. Conserving the Plains: The Soil Conservation Service in the Great Plains. Agricultural History 64(2):58–73.

Hironaka, M., M. A. Forsberg, and H. H. Winward. 1983. Sagebrush-grass Habitat Types of Southern Idaho. University of Idaho Forests Wildlife and Range Experiment Station Bulletin No. 35. Moscow, Idaho: University of Idaho.

Holechek, J. L., R. D. Pieper, and C. H. Herbel. 1989. Range Management: Principles and Practices. Englewood Cliffs, N.J.: Prentice-Hall.

Holling, C. S. 1973. Resilience and stability of ecological systematics. Annual Review of Ecological Systematics 4:1–23.

Hugie, V. K., H. B. Passey, and E. W. Williams. 1964. Soil taxonomic units and potential plant community relationships in a pristine range area of southern Idaho. Pp. 190–204 in Special Publication No. 5. Madison, Wis.: American Society of Agronomy.

Humphrey, R. R. 1947. Range forage evaluation by the range condition method. Journal of Forestry 45:10–16.

International Joint Commission. 1991. A Proposed Framework for Developing Indicators of Ecosystem Health for the Great Lakes Region. Canada: The Commission.

Jardine, J. T., and M. Anderson. 1919. Pp. 16–23 in Range Management on the National Forests. U.S. Department of Agriculture Bulletin 790. Washington, D.C.: U.S. Department of Agriculture.

Johnson, W. R. 1985. Soil Survey of Catron County, New Mexico. Washington, D.C.: Soil Conservation Service, U.S. Department of Agriculture, and Bureau of Land Management, U.S. Department of the Interior, and Las Cruces: New Mexico Agricultural Experiment Station.

Johnson, D. W., C. G. Shaw, and M. Schomaker. 1992. Forest Health Monitoring Plan for Colorado. Technical Report R2. Denver, Colo.: U.S. Department of Agriculture, Forest Service.

Joyce, L. A. 1989. An analysis of the range forage situation in the United States: 1989–2040. General Technical Report RM-190. Fort Collins, Colo.: Rocky Mountain Forest and Range Experiment Station, U.S. Forest Service, U.S. Department of Agriculture.

Klock, G. O. 1982. Some soil erosion effects on forest soil productivity. Pp. S53–S66 in Determinants of Soil Loss Tolerance, D. M. Kral, ed. Special Publication No. 45. Madison, Wis.: American Society of Agronomy.

Klopatek, J. M., R. J. Olson, C. J. Emerson, and J. L. Jones. 1979. Land-use conflicts with natural vegetation in the United States. Environmental Conservation 6(3):191–199.

Lal, R., and B. A. Stewart. 1990. Soil degradation: A global threat. Advances in Soil Science 11:xiii-xvii.

Larsen, L. S., T. G. Barber, E. L. Wesswick, D. E. McCoy, and J. B. Harman. 1966. Soil Survey of Elbert County, Colorado (Eastern Part). Washington, D.C.: Soil Conservation Service, U.S. Department of Agriculture, and Fort Collins: Colorado Agriculture Experiment Station.

Larson, W. E., F. J. Pierce, and R. H. Dowdy. 1983. The threat of soil erosion to long-term crop production. Science 219:458–465.

Lauenroth, W. K. 1985. New directions for rangeland condition analysis. Paper presented at the Symposium on Rangeland Monitoring, Salt Lake City, Utah, February 13, 1985.

Laycock, W. A. 1989. Secondary succession and range condition criteria: Introduction to the problem. Pp. 1–18 in Secondary Succession and the Evaluation of Rangeland Condition, W. K. Lauenroth and W. A. Laycock, eds. Boulder, Colo.: Westview.

Laycock, W. A., and D. A. Price. 1970. Factors Influencing Forage Quality. Pp. 37–47 in Range and Wildlife Habitat Evaluation—A Research Symposium. Miscellaneous Publication No. 1147. Washington, D.C.: U.S. Department of Agriculture, U.S. Forest Service.

Logan, T. J. 1990. Chemical degradation of soil. Advances in Soil Science 11:187–222.

Mabbut, J. A. 1968. Review of concepts of land classification. Pp. 11–28 in Land Evaluation, G. A. Stewart, ed. Australia: MacMillan.

Martin, W. E. 1984. Mitigating the Economic Impacts of Agency Programs for Public Rangelands. Pp. 1659–1680 in Developing Strategies for Rangeland Management: A Report. Boulder, Colo.: Westview.

McClure, J. P., N. D. Cost, and H. A. Knight. 1979. Multiresource inventories—A new concept for forest survey. U.S. Forest Service, U.S. Department of Agriculture, Research Paper Se-101. Asheville, N.C.: Southeastern Forest Experiment Station.

McIntosh, R. P. 1980. The relationship between succession and the recovery process in ecosystems. Pp. 11–62 in The Recovery Process in Damaged Ecosystems, J. Cairns, ed. Ann Arbor, Mich.: Ann Arbor Science Publishers.

Miller, R. W., and R. L. Donahue. 1990. Soils: An Introduction to Soils and Plant Growth, 6th Ed. Englewood Cliffs, N.J.: Prentice-Hall.

Milliron, E. L. 1985. Soil Survey of Cedar County, Nebraska. Washington, D.C.: Soil Conservation Service, U.S. Department of Agriculture, and Lincoln: Conservation and Survey Division, University of Nebraska.

National Research Council of Canada. 1985. The Role of Biochemical Indicators in the Assessment of Aquatic Ecosystem Health: Their Development and Validation. Report No. NRCC 24371. Ottawa: National Research Council of Canada.

Nebel, B. J. 1981. Environmental Science: The Way the World Works. Englewood Cliffs, N.J.: Prentice-Hall.

Odum, E. P. 1969. The strategy of ecosystem development. Science 164:262–270.

Passey, H. B., and V. K. Hugie. 1962. Sagebrush on **relict** ranges in the Snake River Plains and Northern Great Basin. Journal of Range Management 15:273–278.

Passey, H. B., V. K. Hugie, E. W. Williams, and D. E. Ball. 1982. Relationships between soil, plant community, and climate on rangelands of the Intermountain West. Soil Conservation Service Technical Bulletin No. 1669. Washington, D.C.: U.S. Department of Agriculture.

Pastor, J., R. J. Naiman, B. Dervey, and P. McInnes. 1988. Moose, microbes, and the boreal forest. BioScience 38:778–785.

Paul, E. A., and F. E. Clark. 1989. Soil Microbiology and Biochemistry. San Diego: Academic Press.

Paul, R. W. 1988. The Far West and the Great Plains in Transition, 1859–1900. New York: Harper & Row.

Peffer, L. 1951. The Closing of the Public Domain. Stanford, Calif.: Stanford University Press.

Pendleton, D. T. 1989. Range condition as used in the Soil Conservation Service. Pp. 17–34 in Secondary Succession and the Evaluation of Rangeland Condition, W. K. Lauenroth and W. A. Laycock, eds. Boulder, Colo.: Westview.

Pierce, F. J. 1991. Erosion productivity impact prediction. Pp. 35–52 in Soil Management for Sustainability, R. Lal and F. J. Pierce, eds. Ankeny, Iowa: Soil and Water Conservation Society.

Powell, J. W. 1879. Report on the Lands of the Arid Region of the United States. Washington, D.C.: U.S. Government Printing Office.

Rag, B. W., J. B. Fehrenbacher, R. Rehner, and L. C. Acker. 1976. Soil Survey: Stephenson County, Illinois. Urbana: University of Illinois Agricultural Experiment Station, and Washington, D.C.: Soil Conservation Service, U.S. Department of Agriculture.

Rauzi, F., and C. L. Fly. 1968. Water Intake on Midcontinental Rangelands as Influenced by Soil and Plant Cover. Agricultural Research Service and Soil Conservation Service, U.S. Department of Agriculture, and Wyoming Agricultural Experiment Station Technical Bulletin No. 1390. Washington, D.C.: U.S. Government Printing Office.

Retzer, J. L. 1974. Alpine soils. In Arctic and Alpine Environments, J. D. Ives and R. G. Barry, eds. London: Methuen.

Risser, P. G. 1989. Range condition analysis: Past, present and future. Pp. 143–156 in Secondary Succession and the Evaluation of Rangeland Condition, W. K. Lauenroth and W. A. Laycock, eds. Boulder, Colo.: Westview.

Rives, J. L. 1980. Soil Survey of Pecos County, Texas. Soil Conservation Service, U.S. Department of Agriculture, and Texas Agriculture Experiment Station. Washington, D.C.: Soil Conservation Service, U.S. Department of Agriculture.

Rowley, W. D. 1985. U.S. Forest Service Grazing and Rangelands. College Station: Texas A&M University Press.

Royte, E. 1990. Showdown in cattle country. Pp. 60–70 in The New York Times Magazine, December 16, 1990.

Ryan, T. M., and D. Giger. 1988. Soil Survey of Angeles National Forest, California. Washington, D.C.: U.S. Forest Service, U.S. Department of Agriculture, in cooperation with the Soil Conservation Service, U.S. Department of Agriculture.

Sampson, A. W. 1917. Succession as a factor in range management. Journal of Forestry 15:593–596.

Sampson, A. W. 1919. Plant Succession Relation to Range Management. Pp. 1–76 in U.S. Department of Agriculture Technical Bulletin No. 791. Washington, D.C.: U.S. Department of Agriculture.

Sampson, A. W. 1923. Range and Pasture Management. Boston: Wiley.

Sampson, A. W. 1952. Range Management. New York: Wiley.

Sarvis, J. T. 1920. Composition and density of the native vegetation in the vicinity of the Northern Great Plains Field Station. Journal of Agricultural Research 19:63–72.

Sarvis, J. T. 1941. Grazing Investigations on the Northern Great Plains. U.S. Department of Agriculture Technical Bulletin No. 308. Washington, D.C.: U.S. Department of Agriculture.

Satterlund, D. R. 1972. Wildland Watershed Management. New York: Wiley.

Schaeffer, D. J., E. E. Herricks, and H. W. Kerster. 1988. Ecosystem Health, 1: Measuring ecosystem health. Environmental Management 12:445–455.

Sharp, L. A., K. Sanders, and N. Rimbey. 1990. Forty years of change in a shadescale stand in Idaho. Rangelands 12:313–328.

Sharpe, M. 1979. Rangeland Condition. Pp. 29–33 in Proceedings, Symposium on Rangeland Policies for the Future, Tucson, Ariz. Report GTR-WO-17. Washington, D.C.: U.S. Forest Service, U.S. Department of Agriculture.

Shaw, B. 1990. The Lone Ranger Rides Again. People Magazine, December 19, 1990.

Shaxon, T. F., N. W. Hudson, D. W. Sanders, E. Roose and W. C. Moldenhauer. 1989. Land Husbandry: A Framework for Soil & Water Conservation. Ankeny, Iowa: Soil Conservation Society.

Sheridan, D. 1981. Desertification in the United States. Council on Environmental Quality. Washington, D.C.: U.S. Government Printing Office.

Shiflet, T. N. 1973. Range sites and soils in the United States. Pp. 26–33 in Arid Shrublands: Proceedings of the Third Annual Workshop of the United States/Australia Rangeland Panel, D. H. Hyder, ed. Denver: Society for Range Management.

Singer, M. J., and D. N. Munns. 1987. Soils: Introduction. New York: Macmillan.

Smith, J. 1896. Forage conditions of the prairie region. Pp. 309–324 in USDA Yearbook of Agriculture—1895. Washington, D.C.: U.S. Department of Agriculture.

Smith, L. E. 1989. Range condition and secondary succession: A critique. Pp. 103–142 in Secondary Succession and the Evaluation of Rangeland Condition, W. K. Lauenroth and W. A. Laycock, eds. Boulder, Colo.: Westview.

Society for Range Management. 1989. Assessment of Rangeland Condition and Trend of the United States. Denver: Society for Range Management.

Society for Range Management, Range Inventory Standardization Committee. 1983. Guidelines and Terminology for Range Inventories and Monitoring. Denver: Society for Range Management.

Society for Range Management, Task Group on Unity in Concepts and Terminology. 1991. New Directions in Range Condition Assessment. Report to the Board of Directors. Denver: Society for Range Management.

Spence, L. F. 1938. Range management for soil and water conservation. Utah Juniper 9:18–25.

Standing, A. R. 1933. Ratings of forage species for grazing surveys based on volume produced. Utah Juniper 4:11–16, 41–42.

Stoddart, L. A., and A. D. Smith. 1943. Range Management. New York: McGraw-Hill.

Stoddart, L. A., and A. D. Smith. 1955. Range Management, 2nd Ed. New York: McGraw-Hill.

Stoddart, L. A., A. D. Smith, and T. Box. 1975. Range management, 3rd Ed. New York: McGraw-Hill.

Thompson, P. 1989. Poison Runoff: A Guide to State and Local Control of Nonpoint Source Water Pollution. New York: Natural Resources Defense Council.

Thorp, J., T. W. Glassey, T. J. Dunnewald, and B. L. Parsons. 1939. Soil Survey of Sheridan County, Wyoming. Bureau of Chemistry and Soils, U.S. Department of Agriculture. Series 1932, Report 33. Washington, D.C.: U.S. Government Printing Office.

Troeh, F. R., J. A. Hobbs, and R. L. Donahue. 1991. Soil and Water Conservation. Englewood Cliffs, N.J.: Prentice-Hall.

Tueller, P. T. 1973. Secondary succession, disclimax, and range condition standards in desert shrub vegetation. Pp. 57–65 in Proceedings of the Third Workshop of U.S./Australia Rangelands Panel, Tucson, Ariz., D. N. Hyder, ed. Denver: Society for Range Management.

U.S. Congress, Senate. 1936. The Western Range. Senate Document No. 199. 74th Cong., second sess. Washington, D.C.: U.S. Government Printing Office.

U.S. Department of Agriculture, Soil Conservation Service. 1976. National Range Handbook. Washington, D.C.: U.S. Department of Agriculture.

U.S. Department of Agriculture, Soil Conservation Service. 1987. National Resources Inventory. 1987 Instructions for Collecting Sample Data. Washington, D.C.: U.S. Department of Agriculture, and Ames, Iowa: Statistical Laboratory, Iowa State University.

U.S. Department of Agriculture, Soil Conservation Service. 1989a. The Second RCA Appraisal; Soil, Water and Related Resources on Nonfederal Land in the United States; Analysis of Condition and Trends. Washington, D.C.: U.S. Department of Agriculture.

U.S. Department of Agriculture, Soil Conservation Service. 1989b. Summary Report 1987. National Resources Inventory. Statistical Bulletin No. 790. Ames, Iowa: Statistical Laboratory, Iowa State University.

U.S. Department of Agriculture, U.S. Forest Service. 1985. Forest Service Resource Inventory: An Overview. Forest Resources Economics Research Staff. Washington, D.C.: U.S. Department of Agriculture.

U.S. Department of Agriculture, U.S. Forest Service. 1989a. An Analysis of the Land Base Situation in the U.S.: 1989–2040. General Technical Report RM-181. Fort Collins, Colo.: Rocky Mountain Forest and Range Experiment Station, U.S. Forest Service, U.S. Department of Agriculture.

U.S. Department of Agriculture, U.S. Forest Service. 1989b. Resource Planning Act Assessment of the Forest and Rangeland Situation in the United States, 1989. Washington, D.C.: U.S. Department of Agriculture.

U.S. Department of Agriculture, U.S. Forest Service. 1990a. Critique of Land Management Planning. Document FS-455. Policy Analysis Staff. Washington, D.C.: U.S. Department of Agriculture.

U.S. Department of Agriculture, U.S. Forest Service. 1990b. Draft Environmental Impact Statement—Stanley Basin Cattle and Horse Allotment Management Plan, Sawtooth, Idaho, National Forest and National Recreation Area. Washington, D.C.: U.S. Department of Agriculture.

U.S. Department of Agriculture, U.S. Forest Service. 1991a. Draft Environmental Impact Statement—Upper Ruby Cattle and Horse Allotment, Beaverhead, Montana National Forest. Washington, D.C.: U.S. Department of Agriculture.

U.S. Department of Agriculture, U.S. Forest Service. 1991b. Ecological Classification and Inventory Handbook. FSH 2090.11. Washington, D.C.: U.S. Department of Agriculture.

U.S. Department of Agriculture, U.S. Forest Service. 1991c. Forest Service Manual. Washington, D.C.: U.S. Department of Agriculture.

U.S. Department of the Interior. 1979. Desertification in the U.S.: Status and Issues. Working Review Draft. Paris: United Nations Educational, Scientific, and Cultural Organization.

U.S. Department of the Interior, Bureau of Land Management. 1973. Determination of Erosion Condition Class, Form 7310–12. May. Washington, D.C.: U.S. Department of the Interior.

U.S. Department of the Interior, Bureau of Land Management. 1983a. Draft Resource Management Plan and Environmental Impact Statement, Shoshone-Eureka Resource Area, Nevada. Washington, D.C.: U.S. Department of the Interior.

U.S. Department of the Interior, Bureau of Land Management. 1983b. Honey Lake-Beckwourth, California Grazing Program Draft Environmental Impact Statement. Washington, D.C.: U.S. Department of the Interior.

U.S. Department of the Interior, Bureau of Land Management. 1984. 50 Years of Public Land Management. Washington, D.C.: U.S. Department of the Interior.

U.S. Department of the Interior, Bureau of Land Management. 1985a. Bureau of Land Management Manual. Washington, D.C.: U.S. Department of the Interior.

U.S. Department of the Interior, Bureau of Land Management. 1985b. Draft Lemhi, Idaho, Resource Management Plan and Environmental Impact Statement. Washington, D.C.: U.S. Department of the Interior.

U.S. Department of the Interior, Bureau of Land Management. 1985c. Draft Resource Management Plan and Environmental Impact Statement, Walker, Nevada, Planning Area. Washington, D.C.: U.S. Department of the Interior.

U.S. Department of the Interior, Bureau of Land Management. 1987a. National Resources Inventory. 1987 Instructions for Collecting Sample Data. Washington, D.C.: U.S. Department of Agriculture, and Ames: Statistical Laboratory, Iowa State University.

U.S. Department of the Interior, Bureau of Land Management. 1987b. Proposed Resource Management Plan and Final Environmental Impact Statement, San Juan, Utah, Resource Area. Washington, D.C.: U.S. Department of the Interior.

U.S. Department of the Interior, Bureau of Land Management. 1987c. Draft Resource Management Plan and Environmental Impact Statement, Pocatello, Idaho, Resource Area. Washington, D.C.: U.S. Department of the Interior.

U.S. Department of the Interior, Bureau of Land Management. 1989. Public Land Statistics. Washington, D.C.: U.S. Department of the Interior.

U.S. Department of the Interior, Bureau of Land Management. 1990. State of the Public Rangelands: 1990. Washington, D.C.: U.S. Department of the Interior.

U.S. Environmental Protection Agency. 1992. Environmental Monitoring and Assessment Program: 1992 Project Descriptions, Report No. EPA/600/R92/146. Washington, D.C.: U.S. Environmental Protection Agency.

U.S. General Accounting Office. 1988a. More Emphasis Needed on Declining and Overstocked Grazing Allotments. GAO/RCED-88-80. Gaithersburg, Md.: U.S. General Accounting Office.

U.S. General Accounting Office. 1988b. Some Riparian Areas Restored but Widespread Improvement Will Be Slow. GAO/RCED-88-105. Gaithersburg, Md.: U.S. General Accounting Office.

Voss, P. 1960. Politics and Grass. Seattle: University of Washington Press.

Wagner, R. E. 1989. History and development of site and condition criteria in the Bureau of Land Management. Pp. 35–48 in Secondary Succession and the Evaluation of Rangeland Condition, W. K. Lauenroth and W. A. Laycock, eds. Boulder, Colo.: Westview.

Wald, J., and D. Alberswerth. 1985. Our Ailing Rangelands—Condition Report 1985. Washington, D.C.: National Wildlife Federation and Natural Resources Defense Council.

Wald, J., and D. Alberswerth. 1989. Our Ailing Public Rangelands: Still Ailing! Condition Report 1989. Washington, D.C.: National Wildlife Federation and Natural Resources Defense Council.

Wallace, H. A. 1936. The Western Range. Letter from the Secretary of Agriculture, Washington, D.C.

Wasser, C. H. 1977. Early development of technical range management ca. 1895–1945. Agricultural History 51(1):63–77.

Weaver, J. E. 1954. North American Prairie. Lincoln, Nebr.: Johnsen Publishing.

Weaver, J. E., and F. E. Clements. 1938. Plant Ecology, 2nd Ed. New York: McGraw-Hill.

Weaver, J. E., and T. J. Fitzpatrick. 1932. Ecology and relative importance of the dominants of tall-grass prairie. Botanical Gazette 93:113–150.

West, N. E. 1982. Approaches to synecological characterization of wildlands in the intermountain west. Pp. 633–642 in In-place Resource Inventories: Principles and Practices, Proceedings of a Symposium, 1981, T. B. Thomas, L. D. House IV, and H. G. Lund, eds. Washington, D.C.: Society of American Foresters.

West, N. E. 1985. Origin and early development of the range condition and trend concepts. Paper presented at the Symposium on Rangeland Monitoring, Salt Lake City, Utah, February 13, 1985.

Westoby, M. 1980. Elements of a theory of vegetation dynamics in arid rangelands. Israel Journal of Botany 28:169–194.

Westoby, M., B. Walker, and I. Noy-Meir. 1989. Opportunistic management for rangelands not at equilibrium. Journal of Range Management 42:266–274.

Whicker, A. D., and J. K. Detling. 1988. Ecological consequences of prairie dog disturbances. BioScience 38:778–785.

Whitman, R. 1989. Clean water or multiple use? Best management practices for water quality control in the national forest. Ecology Law Quarterly 16:883.

Wight, J. R., and F. H. Siddoway. 1982. Determinants of soil loss tolerance for rangelands. Pp. 67–74 in Determinants of Soil Loss Tolerance, D. M. Kral, ed. Special Publication No. 45. Madison, Wis.: American Society of Agronomy.

Williams, E. W., and V. K. Hugie. 1966. Range soil studies. Soil Conservation 31:147–149.

Wilson, A. D. 1989. The development of systems of assessing the condition of rangeland in Australia. Pp. 77–102 in Secondary Succession and the Evaluation of Rangeland Condition, W. K. Lauenroth and W. A. Laycock, eds. Boulder, Colo.: Westview.

Wilson, A. D., and G. J. Tupper. 1982. Concepts and factors applicable to the measurement of range condition. Journal of Range Management 35:684–689.

Wissel, C. 1984. A universal law of the characteristic return time near thresholds. Oecologia 65:101–107.

Wooldridge, D. D. 1963. Soil properties related to erosion of wild-land soils in central Washington. In Forest-Soil Relationships in North America, C. T. Youngberg, ed. Corvallis: Oregon State University Press.

Wuerthner, G. 1991. How the West was eaten. Wilderness 54:28–37.

Appendix

PLANTS		Present %	Climax %
	Rothrock grama	9	9
	Bush muhly	9	9
	Santa Rita threeawn	2	2
Grass	Spike dropseed	2	2
	Arizona cottontop	2	2
&	Lehmon lovegrass	7	0
	Mesa threeawn	T	
Grass	Sixweek threeawn	2	2
	Annual threeawn	4	4
Like	Needle grama	10	5
47 %			
	Perennial spurge	2	2
	Arrow leaf		
	Lanceleaf ditaxis		
	Spiny goldenhead	3	3
Forbs	Desert marigold		
	Indian wheat		
	Ann. spiderling		
9 %	Eriastrum		
	Pick me not	4	4
	Wildcarrot		
	Fansy mustard		
	Honeymat		
	Combur		
	Burroweed	10	5
	Shortleaf baccharis	1	1
Shrubs	Mormon tea	2	1
	Desert hackberry	6	2
&	Wolfberry	1	1
	Staghorn nicholla	13	
Trees	Jumpino cholla	T	
	Eng. prickley pean	2	5
Rep.	Barrel cactus	1	
	Mesquite	3	3
	Blue paloverde	2	2
42 %	Pencil cholla	T	
	Soapweed	T	

RANGE CONDITION WRITE-UP SHEET

Ranch: SRER
Technician: D. Robinett
Date: 4789 Field# 5N T 18S R 14E Sec. 35
Elevation 3050 Ppt. 11 inches

TREND INDICATORS

Vigor of Key
Forage Plants:

	Rating		
	High	Med	Low
Rothrock grama		x	
Santa Rita threeawn		x	
Bush muhly			x

Seedlings and
Young Key Forage Plants:

	Abun	Some	None
Rothrock grama	x		
Santa Rita threeawn		x	
Bush muhly			x

,Residues and litter:
        Abun   Some   None
Gully-Rill Stabilization:
        Complete Some None
Final Trend Rating:
        Up Sl. Up Static
        Sl. Down   Down
    ESTIMATED FORAGE YIELD
Use Adjustment
Wildlife: __% Tot. Ann. Lbs/Ac: 451
Slope: __% Useable Lbs/Ac: _____
Rocks: __% Useable Lbs/Ac
Brush: __%   @ P.G.U.: _____
Soil & Erosion: __% Ac/An.
Water: __%   Unit Mo.:___
Total: __%   Adjusted: _____

Idaho fescu (*Festuca idahoensis*)

169

Series: Hayhook

Slope: 1-5   Exposure: level

CONDITION INDICATORS

Condition Class Based on Vegetation:

                EC   GC   FC   PC

Active Erosion: Wind Water Sheet Rill
                 Gully
                 None   Slight
                 Moderate Severe
Final Condition Class Rating: 64

Major Land Resource Area 40-1
Range Site Name Deep sandyloam
Write-Up No. 4

About the Authors

JOHN C. BUCKHOUSE is a professor in the Department of Rangeland Resources, Oregon State University. He has received both M.S. and Ph.D. degrees in range science/watershed management from Utah State University. The focus of the majority of his research has been on watershed management and water resources research.

FRANK E. BUSBY (*Chair*) is director of U.S. Programs, Winrock International Institute for Agricultural Development, Morrilton, Arkansas. He holds a Ph.D. in range watershed science from Utah State University. He has had extensive experience in range management, having served as an extension rangeland management specialist, director of the Wyoming Cooperative Extension Service, and associate dean of the College of Agriculture and head of the Department of Range Management, University of Wyoming.

DONALD C. CLANTON received his Ph.D. in animal nutrition from Utah State University. Now retired, he was a professor of animal science at the University of Nebraska. His research includes ruminant nutrition, particularly as it effects reproduction, and range nutrition.

GEORGE C. COGGINS earned a J.D. at the University of Michigan Law School and is currently the Frank Edwards Tyler Distinguished Professor of Law at the University of Kansas Law School. During his career he has focused on public land law and natural resource law.

GARY R. EVANS is the chief science advisor to the Secretary of Agriculture and the staff for global change issues, U.S. Department of Agriculture, Arlington, Virginia. He holds a Ph.D. in natural resources decision theory. His primary research has been in forest biology and ecology.

Antelope bitterbrush (*Purshia tridentata*)

KIRK L. GADZIA is president of Resource Management Services, Bernalillo, New Mexico. He holds an M.S. in range science. His career focus has been on private education of resource managers in the field of rangeland and wildlife management.

CHARLES M. JARECKI has been the owner/operator of a cattle ranch in Polson, Montana, for many years. He attended New York State College of Agriculture and received his degree from Montana State University. His active involvement in range management entails practical and applied research.

LINDA A. JOYCE is project leader at the Rocky Mountain Forest and Range Experiment Station, U.S. Forest Service, Fort Collins, Colorado. Her primary research focus is ecosystem processes on rangelands. She earned her Ph.D. in range ecology at Colorado State University.

DICK LOPER is president and owner of Prairie Winds Consulting Service, Lander, Wyoming, a range management and range science advisory concern. His M.S. in range science is from Kansas State University. As a rangeland specialist, his primary area of research has been natural resource management.

DANIEL L. MERKEL is range conservationist for the Soil Conservation Service, U.S. Department of Agriculture, Denver, Colorado. His research work includes land classification, renewable resource inventory techniques, range improvement and management practices, and critical area stabilization. He has focused on soil science and range management.

GEORGE B. RUYLE is associate research scientist and range management extension specialist, School of Renewable and Natural Resources, University of Arizona. His research interests include plant and animal interactions on rangelands. He earned his Ph.D. in range science from Utah State University.

JACK WARD THOMAS is chief research wildlife biologist and project leader of the Range and Wildlife Habitat Reserve, Pacific Northwest Forest and Range Experiment Station, U.S. Forest Service, LaGrande, Oregon. He received his Ph.D. from the University of Massachusetts in wildlife biology. His research includes impact of wildlife populations in wild and urbanized settings, sociobioeconomic implications of game habitat manipulation, and forestry/wildlife relationships.

JOHANNA H. WALD is a senior staff attorney with the Natural Resources Defense Council in its Washington, D.C., office. She received her LL.B. at Yale Law School and has focused during her career on federal land and resource management and environmental law. Her expertise is in federal range management issues.

STEPHEN E. WILLIAMS is a professor and department head at the University of Wyoming. His teaching and research interests include soil biology and biochemistry, forest and range soils, and soil microbial ecology. He earned his Ph.D. in soil science and microbiology at North Carolina State University.

Index

Big bluestem (*Andropogon gerardii*)

Recent Publications of the Board on Agriculture

Policy and Resources

Soil and Water Quality: An Agenda for Agriculture (1993), 510 pp., ISBN 0-309-04933-4

Managing Global Genetic Resources: Agricultural Crop Issues and Policies (1993), 450 pp., ISBN 0-309-04430-8

Pesticides in the Diets of Infants and Children (1993), 408 pp., ISBN 0-309-04875-3

Managing Global Genetic Resources: Livestock (1993), 294 pp., ISBN 0-309-04394-8

Sustainable Agriculture and the Environment in the Humid Tropics (1993), 720 pp., ISBN 0-309-04749-8

Agriculture and the Undergraduate: Proceedings (1992), 296 pp., ISBN 0-309-04682-3

Water Transfers in the West: Efficiency, Equity, and the Environment (1992), 320 pp., ISBN 0-309-04528-2

Managing Global Genetic Resources: Forest Trees (1991), 244 pp., ISBN 0-309-04034-5

Managing Global Genetic Resources: The U.S. National Plant Germplasm System (1991), 198 pp., ISBN 0-309-04390-5

Sustainable Agriculture Research and Education in the Field: A Proceedings (1991), 448 pp., ISBN 0-309-04578-9

Toward Sustainability: A Plan for Collaborative Research on Agriculture and Natural Resource Management (1991), 164 pp., ISBN 0-309-04540-1

Investing in Research: A Proposal to Strengthen the Agricultural, Food, and Environmental System (1989), 156 pp., ISBN 0-309-04127-9

Alternative Agriculture (1989), 464 pp., ISBN 0-309-03985-1

Understanding Agriculture: New Directions for Education (1988), 80 pp., ISBN 0-309-03936-3

Designing Foods: Animal Product Options in the Marketplace (1988), 394 pp., ISBN 0-309-03798-0; ISBN 0-309-03795-6 (pbk)

Agricultural Biotechnology: Strategies for National Competitiveness (1987), 224 pp., ISBN 0-309-03745-X

Regulating Pesticides in Food: The Delaney Paradox (1987), 288 pp., ISBN 0-309-03746-8

Pesticide Resistance: Strategies and Tactics for Management (1986), 480 pp., ISBN 0-309-03627-5

Pesticides and Groundwater Quality: Issues and Problems in Four States (1986), 136 pp., ISBN 0-309-03676-3

Soil Conservation: Assessing the National Resources Inventory, Volume 1 (1986), 134 pp., ISBN 0-309-03649-9; Volume 2 (1986), 314 pp., ISBN 0-309-03675-5

New Directions for Biosciences Research in Agriculture: High-Reward Opportunities (1985), 122 pp., ISBN 0-309-03542-2

Genetic Engineering of Plants: Agricultural Research Opportunities and Policy Concerns (1984), 96 pp., ISBN 0-309-03434-5

continued

Rangeland health

Nutrient Requirements of Domestic Animals Series and Related Titles

Nutrient Requirements of Fish (1993), 108 pp., ISBN 0-309-04891-5

Nutrient Requirements of Horses, Fifth Revised Edition (1989), 128 pp., ISBN 0-309-03989-4; diskette included

Nutrient Requirements of Dairy Cattle, Sixth Revised Edition, Update 1989 (1989), 168 pp., ISBN 0-309-03826-X; diskette included

Nutrient Requirements of Swine, Ninth Revised Edition (1988), 96 pp., ISBN 0-309-03779-4

Vitamin Tolerance of Animals (1987), 105 pp., ISBN 0-309-03728-X

Predicting Feed Intake of Food-Producing Animals (1986), 95 pp., ISBN 0-309-03695-X

Nutrient Requirements of Cats, Revised Edition (1986), 87 pp., ISBN 0-309-03682-8

Nutrient Requirements of Dogs, Revised Edition (1985), 79 pp., ISBN 0-309-03496-5

Nutrient Requirements of Sheep, Sixth Revised Edition (1985), 106 pp., ISBN 0-309-03596-1

Nutrient Requirements of Beef Cattle, Sixth Revised Edition (1984), 90 pp., ISBN 0-309-03447-7

Nutrient Requirements of Poultry, Eighth Revised Edition (1984), 71 pp., ISBN 0-309-03486-8

Further information, additional titles (prior to 1984), and prices are available from the National Academy Press, 2101 Constitution Avenue, NW, Washington, DC 20418, 202/334-3313 (information only); 800/624-6242 (orders only); 202/334-2451 (fax).